The Court Martial of Apache Kid

Based on the Original Transcript

Victoria Sutton, MPA, PhD, JD

ISBN: 978-0-9968186-4-3

Published by Vargas Publishing
P.O. Box 6801
Lubbock, TX 79493

Acknowledgements

This work is the product of a lot of people over a period of more than one hundred years. With regard to the transcript of the court martial trial, my thanks go to the court reporters who obviously took great care in transcribing the proceedings of the court martial, and to the officers who served on the court. The translators are also to be given credit for ensuring that three different languages turned into one English document. The archivists from the U.S. Army Judge Advocates to the archivists at the United States Archives all had to do a flawless job to have this transcript discoverable after lying unpublished for 140 years. Although I read from historians that the transcript was lost to time, or never existed, my law librarian, Jamie Baker, did not accept that answer. Following a citation to the U.S. Archives in Washington, D.C., I found an enthusiastic archivist there who would not give up, going from one index to another and to another. At the end of that trail, safely in the bowels of the U.S. Archive Building in Washington, D.C. still on the yellow-lined paper filled with fountain-pen skilled writing, was the original, hand-written transcript. Thanks to all of the individuals and the institutions they represent for the care they took along each step of this journey in preserving these historical legal documents.

After I had completed transcribing the original documents, I sought the military law expertise from my colleagues, former Judge Advocate, Col. Richard Rosen (ret.); and Judge Advocate General Walter Huffman (ret.). They are the co-authors of the leading military law casebook, and are among the most respected authorities on military law in the nation. Through them, I was also directed to the Judge Advocate General Historian, Frederick Borch, III, who is a wealth of knowledge and expertise on the history of the Judge Advocate General.

This was a journey that was intended to give a perspective and a voice to the San Carlos Apache in this historical event which led to Apache Kid becoming a western American legend. Getting to know who he was, was more difficult than I had even imagined, but with the help of Twila Cassadore, Ramon Riley, Anthony Logan, Sadie Kniffen, Seth Pilsk and Cornelia Bush and many others, I was able to bring some of their perspectives to this story. Thank you.

Acknowledgements

Table of Contents

Introduction

I started research on this project in 2010 when I read about the court martial case of Apache Kid, but found there was no citation to the trial or the trial transcript. It was 1887, and it was not surprising that it might have disappeared. Even an author of one of the more recent historical books on the court martial of the Apache Kid, Clare V. McKanna, Jr. , said in a radio interview that the transcript did not exist.[1] Having done a good bit of historical legal research I thought I would at least look for it. The historian for the U.S. Army Judge Advocate 's office suggested that these records were no longer in the possession of the U.S. Army and would have likely been sent to the U.S. Archives. Enlisting the help of my law librarian, Jamie Baker; she searched archived resources and found a citation which located the trial transcript in the U.S. Archives. She contacted the U.S. Archives but discovered the transcript was not available for sharing. Any chance of retrieving these documents would have to be in person, at the U.S. Archives building in Washington, D.C.

In many ways, finding the transcript was really the beginning of the journey, not the end. I began the transcription process as soon as I returned, and that was a journey in itself, as I often anxiously watched the trial unfold deep into the night. After several months, I completed the work of transcribing the handwritten documents in early 2017.

First, I was fortunate to have military law colleagues who are among the most respected and knowledgeable in their field of military law. I asked them questions I had about the transcript and eventually drew heavily on their expertise to determine what was unusual and what might have been normal procedure in the trial.

I started my search for descendants of Apache Kid, and started with the White Mountain Apache tribe. After speaking to a very knowledgeable White Mountain Apache cultural expert, Ramon Riley, I learned that although most of the U.S. Army Scouts came from this tribe, the Apache Kid actually did not. He was from the San Carlos "band" of Indians, the area now

---

[1] SDSU history professor Clare McKanna talks about the plight of Native Americans in his new book "Court-Martial of Apache Kid, Renegade of Renegades." Aug 19, 2009 at http://www.kpbs.org/news/2009/aug/19/sdsu-professor-chronicles-apache-kid/ (last visited May 28, 2018).

designated as the San Carlos Apache Reservation. I shifted my focus to the south, and with a few connections made, I was able to reach incredibly knowledgeable people at San Carlos. Some families knew about Apache Kid, others did not. It would become clearer why that was so as I continued to talk to people. As you might expect, the Apaches who joined with the U.S. Army found themselves facing members of their own band in enforcement actions, causing a divide among these families. Although the families have moved past their former conflicts, in general, they still remember their heritage through oral histories and songs.

Context for the Charges of Desertion and Mutiny

It was 1887, and relations between the United States and tribes were changing rapidly, where the President was no longer signing treaties with tribes[2] and the wars against America's native people had almost completely ended from the rush of Manifest Destiny that crushed the tribes that did not move out of the way.[3]

It was only in 1878, that the U.S. Army instituted the requirement that court martial proceedings must be transcribed. It was also a requirement that any death penalty sentence had to be approved by the President of the United States.

Such a case had occurred in a tragic encounter, the Massacre at Cibecue. The Apache Army Scouts in August 1881 had technically mutinied against their commanding officers when their medicine man, Dancer, was arrested, but instead of taking him into custody, they killed him in a horrific manner. The Army feared that Dancer would influence the Apaches to revolt, although there was no direct evidence of that occurring. The Apache Scouts then attacked the commanding officers killing a number of them, and escaping into the mountains. Several were captured (Dandy Jim, Deadshot, Skippy and Mucheco) and tried in a court martial at Fort Grant in November 1881. Mucheco received a life sentence in Alcatraz. Dandy Jim, Deadshot and Skippy were given death sentences. Pres. Chester Arthur (not elected, but taking office after the assassination of Pres. Garfield) approved their death sentences. The three scouts were hanged at Fort Grant, March 3, 1882.

---

[2] In 1871, Congress ended the President's treatymaking authority: "Provided, that hereafter no Indian nation or tribe within the territory of the United States shall be acknowledged or recognized as an independent nation, tribe, or power with whom the United States may contract by treaty … ." 16 Stat 544, 566 (Mar 3, 1871). The legislative history showed the diminished power and landholdings of tribes was part of the reason Congress no longer wanted to recognize tribes as nations. See David P. Currie, Indian Treaties, 10 GREEN BAG 2D 445 (Summer 2007). I argue this is plainly unconstitutional on its face, but that was the state of federal Indian policy in 1871.

[3] Academics consider the Nez Pearce War of 1877 to be the last great conflict; others consider the Battle of Wounded Knew of 1891 to be the last Indian War. But the battles in the West were just as fierce although smaller and these were still occurring during the period of this court martial case. The Battle of Cibecue in 1881 between the U.S. Army and two bands of Apaches is a battle that is still remembered by the descendants and others, today. It likely had an influence on the Apache Kid's court martial case, just six years later, because of the criticism the U.S. Army received for its lack of due process in the executions of Apaches with no trial at the end of the battle.

Al Sieber's Apache Scouts escaped to the top of a hill and defended the Cavalry, and the Apache Kid could have been among them. Yet, if so, this show of loyalty was not enough to prevent the charge of mutiny, despite the lack of evidence.

The first battle in which Apache Kid is identified in the literature is in July 1882 at Big Dry Wash.[4]

In 1883, the Arizona territorial legislature petitioned to eliminate all Indian reservations, and specifically in 1885, they petitioned to eliminate both the San Carlos Reservation and the White Mountain Reservation.[5]

It was at one of the tizwin (alcohol) gatherings, December 25, 1886, that the Apache Kid's father or grandfather was killed. Sieber sent Apache Kid with some other Apache Scouts to break up the gathering, and Apache Kid came upon the murder of his father or grandfather. (The word for father and grandfather in Apache is very similar and the record is not clear about whether it was his father or grandfather.)

A major earthquake occurred May 3, 1887, just after the birth of Apache Kid's daughter. This earthquake killed fifty people in Sonora and was seen as a dark omen among the Apache and kept the entire community on edge. The Apache people appealed to Usen for deliverance from the omen, and tizwin gatherings continued to be held to make these appeals.

It is within this unsettling context of bias, hatred, inhumanity, greed and natural disasters that turned what would be a simple absent without leave charge against the Apache Kid into a charge of desertion and mutiny with a sentence of the death penalty at stake, without the requisite evidence to support these crimes.

Reading the transcription

The transcript on the following pages is annotated and footnoted. In the transcript, I have made notations which are highlighted to show where I have added to the transcript to show the name of the speaker, which was mentioned only once at the beginning of a witness's testimony. I have also noted where the transcriptionist changed, evidenced by the handwriting changes.

The page number of the original transcript is in bold at the beginning of the transcribed page. "Page 34", for example, appears at the bottom or top of each page of the original transcript. You will see "Page 34 (3690)". The number in parentheses indicates the photograph number of the photographed page of the transcript cross-referenced to the original photograph of the page from the transcript or other documents.

To the extent the text is recognizable for reading for meaning, the original spelling, misspelling or abbreviations used in the transcript are maintained in the transcription. Also noted,

---

[4] Clare V. McKanna, Jr., Renegade of Renegades 13 (2012).

[5] Paul Andrew Hutton, *The Apache Wars*, 378 (2016).

are the changes in court reporters, when the handwriting changed in the transcript. Also noted are places where the Judge Advocate made edits to the transcript.

Annotations from Judge Advocate General Walter Huffman (ret.) and Judge Advocate Col. Richard Rosen (ret.) who have added remarks and comments that involve both procedural and substantive law. Both are military law experts, authors of the leading law casebook on Military Law and distinguished professors and former Dean and Interim Dean, respectively, at Texas Tech University School of Law. Prof. Rosen is also Director of the Texas Tech University School of Law Military Law Center. They are also my friends and colleagues. You will also see annotations from me, Victoria Sutton.

In this nine day trial, my hope is that you will feel the layers of context in the transcript as you read it, that will put you in the trial and you can not only read it, but experience it.

The Documentary

A documentary of the case based on this original trial transcript was shot in 2017 and completed in 2018. The documentary was shot at the National Ranching Heritage Center, Texas Tech University in Lubbock, Texas. The actor cast in the role of the Apache Kid is descendant of Apache Kid from San Carlos, Arizona.

This original transcript with annotations is a helpful companion document for watching the documentary.

The Transcript[6]

---

[6] "General Court Martial of Apache Kid, Company A, Indian Scouts," Records Received, 4475, December 27, 1887, Judge Advocate General (Army), RG 153, National Archives.

Case 1

[Handwriting Court Reporter no. 1]
Proceedings of a General Court martial convened at San Carlos, Arizona by virtue of the following order:

       Hdqrrs. Dept of Arizona
In the Field
San Carlos, A.T.
June 25th 1887

Field Orders }
No. 1   }

     A General Court Martial is hereby ordered[7] to convene at San Carlos, A.T. at 2 P.M. today or as soon thereafter as practicable for the trial of such persons as may be properly brought before it.
Detail for the Court:
1. Capt. A.H. Bowman, 9th Infantry
2. Capt. R.G. Smither, 10th Cavalry
3. 1st Lt. H. DeLany, 9th Infantry
4. 2d Lieut. C.P. Johnson, 10th Cavalry
5. 2nd Lieut. J.B. Hughes, 10th Cavalry
6. 2nd Lieut. W.G. Elliot, 9th Infantry
   2nd Lieut. L.D. Tyson, 9th Infantry, is appointed Judge advocate.

A greater number of officers than these can not be assembled with

out manifest injury to the service.[8] The Court is authorized to sit without regard to hours.

---

[7] Court-Martial is unlike an Art. III court. These are not standing courts and they only exist when convened. [Huffman]

[8] In 1874, general courts martial had to consist of between 5 and 13 members and actually 13 members was the preferred number when that number could be convened without manifest injury to the service. At the time the trial counsel talked about the members he indicated that the number that was ultimately established by the convening counsel in this case, it had to be a number smaller than 13 or it would cause manifest injury to the service. If the number went below 5 members, you wouldn't have a sufficient quorum, a jurisdictional error in the court martial proceeding. Today, general courts have to consist of at least 5 members, there could be more. In capital cases there have to be at least 12 members. Under a new revision under the Military Justice Act, in 2016, for a general court martial the number has been fixed at 8 members and of course, 12 members are still required for capital cases.[Rosen]

By Command of
    Brig. General Nelson A. Miles
    (Sgnd) Chas. B. Gatewood
    1s Lieut., 6th Cavalry
    Aide-de-Camp

DAY ONE
San Carlos, Arizona
June 28th, 1887

The Court met pursuant to the foregoing order at 9 oclock, and present:
1. Capt. A.H. Bowman, 9th Infantry
2. Capt. R.G. Smither, 10th Cavalry
3. 1st Lieut. H. DeLany, 9th Infantry
4. 2d Lieut. C.P. Johnson, 10th Cavalry
5. 2nd Lieut. J.B. Hughes, 10th Cavalry
6. 2nd Lieut. W.G. Elliot, 9th Infantry
   2nd Lieut. L.D. Tyson, 9th Infantry, is appointed Judge advocate.

[CHANGE IN HANDWRITING Court reporter No. 2 (backhanded slant]

The Court has proceeded to the trial of 1st Seagt. Kid, Co "A" "Indian Scout" who has come before the Court and heard the order Convening read.

Robert McIntosh was duly sworn by the Judge Advocate as interpreter. "Shago' was also duly sworn by the Judge Advocate as Interpreter. 1st Lieut. Jno A. Baldwin, 9th Inf. Appointed as Counsel for the accused[9][10]

Page 3 (3655)
And was regularly introduced before the Court.

The accused objected [Baldwin]: to Capt. A. H. Bowman, 9th Inf. On the ground of having formed and represented an opinion concerning the merits of this case is a witness and is prejudicial against the defendant and asks that he be put on his voire dire

---

[9] There was no right to defense counsel in in the 19th century, in fact, it did not become a statutory right until 1916. Defense counsel could be provided and it could be a civilian although that was frowned upon. It was frowned upon by the military to have any lawyers involved in the case, William Tecumseh Sherman saying just a few years earlier that lawyers just gum up the process paraphrasing his terms. But defense counsel could be appointed and in this case defense counsel was appointed. Defense counsel did not have to be a lawyer and to my knowledge Lieutenant Baldwin was not a lawyer and neither was the trial counsel. [Rosen]

[10] Until the Uniform Code of Military Justice (UCMJ) was enacted after WWII, it was common for both Prosecutor and Defense Counsel to be non-lawyers. [Huffman]

for the purpose of an examination. Capt. Bowman was then duly sworn on his voir dire by the Judge Advocate and testified as follows:

Question by accused [Baldwin]: Have you formed or expressed any opinion concerning the acts or conduct of the accused?

Answer [Bowman]: I have, I was a witness to what occurred here on the 1st of June. I refer to the firing upon the Commanding Officer Al Sieber, Chief of Scouts.

2. by Acc. [Baldwin] Have you any bias, prejudice or opinion that in any way operates against the accused?
Answer [Bowman]: No. As a member of the Court I have not. When I come to be sworn my oath requires me to well and truly try [the accused?] according to the evidence.

The Court was then cleared and closed. The challenged member retiring and after due deliberation the Court was re-opened, the accused and challenged member being present. And the decision of the Court was announced by the Judge Advocate that the challenge is not sustained. The accused then objected to Capt. R.J. Smith for the same reasons that he had objected to Capt. Bowman with the exception that he is not a witness. At the request of the accused Capt. Smith was then duly sworn on his voire dire by the Judge Advocate and testified as follows:

2. by Acc. [Baldwin] Have you any bias, prejudice or opinion in any way detrimental to the interests of the accused, and have you [all?]

Page 4 (3656)
Any bias, expressed any opinion concerning the guilt or innocence of the accused? Have you formed any opinion relating to the alleged acts or conduct of the accused?

Answer [Smith]: I have some biased opinions as to the acts of the accused. As to expression of opinion I am not positive whether I have or not, but I do not think I have said anything particularly prejudiced to the accused. But I am satisfied that I have no feeling nor have I said anything that would present me doing the accused justice on any evidence that may be addressed [?] before this Court. It is true that I said at a distance what occurred on June 1st but I did not know who the parties were until told afterwards. I am ignorant as to who actually did the firing in this incident [?]. I am satisfied that I have not formed any opinion that would present me from doing justice to the accused.

2. by Acc. [Baldwin]: Do your bias and opinions relate to the occurrences of June 1, 1887?

Answer [Smith]: Yes. I relate to what occurred June 1st.

Accused [Baldwin], do you believe that, in ascertaining a verdict, your biased opinions would in no way effect or influence your judgment on the means of evidence that you might determine upon?

Ans. [Smith]: I am satisfied that they would not. Since I am detained as a member of this Court I have fully determined to remoe from my mind or thoughts anything that had previously occurred and be guided solely on the evidence that might be submitted before the Court. I believe I have so expressed myself. The Court was then cleared and closed for deliberation. The challenged member retiring, and after due deliberation the Court was again convened, the challenged member and the accused being present and the Judge

Page 5 (3657),
Advocate announced that the objection had not been sustained by the Court.

The accused [Baldwin] then objected to 1st Lieut. H. DeLany, 9th Infantry, on the following grounds:

Objections. That the member is biased and prejudiced agasint the accused is an important witness in this case, has formed and expressed opinions concerning the guilt of the accused, and has expressed an opinion as to what punishment the accused should receive. At the request of the accused Lt. DeLany was then put on his voire dire and having been duly sworn by the Judge Advocate testified as follows:
2. by Acc. [Baldwin]: Have you any bias feeling or prejudice against the accused for his alleged acts? Have you at any time formed or expressed an opinion relating to the guilt or innocence of the accused?
3. Have you at any time said what should be done with the accused when caught?

Ans. 1 [DeLany] I have. At the same time I have heard no evidence and am perfectly able to try the accused according to my oath.

2. by Acc. [Baldwin]: You witnessed the occurrence at June 1, 1887, were your feelings [m...d] at the persons or parties to that affair?

Ans. [DeLany]: Yes. I do not know of my own knowledge heard [?] that the accused was a party to that affair.

2. by Acc. [Baldwin]: Would not your evidence which you may be called upon to give confirm your bias against the accused and thereby effect your trying this case influenced by any feeling, bias or prejudice ?

Ans. [DeLany]: I think not.

Page 6 (3660)

2. by Acc. [Baldwin]: Can you express an opinion as to what should be done with the accused and divest your mind of feeling, bias or prejudice and be guided solely and clearly by the evidence which you may or may not believe to be true?

Ans. [DeLany]: I can express an opinion as to what should be done with the accused provided he is found guilty and divest myself of all feeling, bias or prejudice and be guided solely by the evidence.

2. by Acc. [Baldwin]: You say you have expressed an opinion as to what should be done with the accused; was that detrimental to the accused and was it of an extreme nature?

Ans. [DeLany]: I do not know whether you can call justice to an accused detrimental or not. It was of an extreme nature.

2. by Acc. [Baldwin]: You are aware that under a member's oath when any doubt arises, he determines, according to his understanding and therefore if at any time you have expressed opinions inimical to the accused would not that doubt be influenced by previous opinions?

Ans. [DeLany]: I can only answer this question by saying that I can try this case according to my oath.

The court was then cleared and closed for deliberation the challenged member retiring and after due deliberation the Court was reconvened the accused and challenged members being present and the decision of the Court was announced by the Judge Advocate that the challenge is not sustained.

The accused [Baldwin] then objected to St. Johnson on the following grounds: on the ground of prejudice because the member has already formed an opinion as to the guilt of the accused. He is an important witness. Has expressed an opinion to the nature of the punishment the accused should receive and that previous to

Page 7 (3661)
Being detailed on this Court the member has assessed the guilt of the accused and therefore his judgment is effected and maliciously {?} formed.

The Counsel for the accused [Baldwin] then requested to have Lieut. Johnson put on his voire dire and Lieut. Johnson having been duly sworn by the Judge Advocate testified as follows:

2. by Acc. [Baldwin] Have you expressed at any time any opinons relative to the accused acts or conduct of the accused?

Ans. [Johnson] This being the first time that I saw the accused I could not have formed a personal prejudice against the accused. In regard to forming an opinion upon the alleged acts of said accused if evidence was plead before this court sustaining ___ [public?] rumor which has been circulated in regard to the alleged acts, my opinion might here be formed by said evidence. At present the opinion in regard to the guilt of the accused is held in abeyance for want of evidence from which to form an opinion in regard to his guilt or innocence.

2. by Acc. [Baldwin] Are you in any way biased, prejudice or inimical against the acused?

Ans. [Johnson] No.

2. by Acc. [Baldwin] Are you the officer who [recully] pursued and rove a party of Indians back to this reservation?

Ans. [Johnson] I am one of the officers.

2. by Acc. [Baldwin] By common notoriety and official information did you not know who were supposed to have been engine in the occurance on June 1st, 1887?

Ans. [Johnson] I had heard the name of the indian who was supposed to have been engaged in a disturbance at San Carlos and I supposed the band that I was following to consisted of the same Indians. As I never feel Commandian with any member of this band it is

Page 8 (3662)
Impossible for me to say whether they were the same or not.

2. by Acc. [Baldwin]Did you at any time previous to being detailed on this Court form or express an opinion concerning these persons?

Ans. [Johnson] Yes. I formed an opinion relative to this band of Indians on several occasions.

2. by Acc. [Baldwin] Did you not check and refer to this band of Indians as having ___ to members indian known as the Kid?

Ans. [Johnson] I, of course, supposed the man called "the Kid" was in this band of Indians. I may have expressed an opinion on several occasions to that effect.

2. by Acc. [Baldwin] Give the nature of the opinion you expressed relative to this band of Indians.

Ans. [Johnson] If I was to give the nature of the many opinions I have expressed about this band of Indians in 17 days trailing, I would delay the Court uselessly.

2. by Acc. [Baldwin] From official information or utterance did you not know that some scouts from San Carlos were among the renegade Indians you were following?

Ans. [Johnson] Yes. I was so informed by official telegram.

2. by Acc. [Baldwin] What opinions did you express or form concerning the conduct or acts of certain Indians at San Carlos who had left their reservation and who you followed?

Ans. [Johnson] Referring to the Scouts in the previous question I was of the opinion occurences such so were commonly reported has really occurred at San Carlos and participated in by these scouts that they were amenable to military Laws and the custom at War like cares with soldiers and I think expressed some such opinions.

2. by Acc. [Baldwin] At any time were your opinions, feelings or prejudices unfriendly to the accused –"the Kid"--- or those Indians previously mentioned?

Page 9 (3664)
Ans. [Johnson] All officers following a public enemy can scarcely be supposed to be friendly to him provided he proposes to do his duty when he overtakes him . But such feeling might mainly cease to exist as it did in my case when the accused voluntarily surrendered without having caused loss of life or injury to myself or Command.

2. by Acc. [Baldwin] Did you form or express any opinion as to what punishment they should receive?

Ans. [Johnson] I did.

2. by Acc. [Baldwin] Did you say what you would do with them if you caught up with them?

Ans. [Johnson] I did.

2. by Acc. [Baldwin] You have stated, your opinions are held in abeyance, and you say that your mind is entirely free of all bias or prejudice against the accused?

Ans. [Johnson] I have stated that all my opinions have been based on rumors more or less authentic and not on facts such as would prevent me from rendering a verdict on such evidence as may be placed before this Court, and to the best of my knowledge and belief I can conceive of no prejudice against the accused in person.

The Court was then cleared and closed for deliberation the challenged member retiring.

The Court was then opened the member and accused returning [to] their seats. The Judge Advocate announced that the objection was not sustained by the Court.

The Court then took a recess until V o'clock P.M.

The Court at V P.M. again proceeded to business. All the members being present and Judge Advocate also presents. The accused and counsel also being present.

The accused then objected to Lieut. J.B. Hughes, based on the following grounds:

Page 10 (3665)
That the member has formed or expressed opinions concerning this case and the acts alleged against the accused. That he is biased and prejudiced against the accused. That he was present at the agency at the time of the alleged occurrences and that finally he is a party to the transactions that may be called upon to judge of facts to which he may testify. At the request of Counsel for the accused the challenged member was there put in his voire dire and being duly sworn by the Judge Advocate testified as follows:

2. by Acc. [Baldwin] Have you any bias or prejudice for or against this defendant?

Ans. [Hughes] No.

2. by Acc. [Baldwin] In your mind at the present time uninfluenced by any feeling for or against the accused and have you any knowldege of any acts relating to the accused or those supposed to have been associated with him from that could in any way effect your opinions or judgement?

Ans. [Hughes]. My mind is uninfluenced against the accused and I have no knowledge of any acts that might make me form any opinion other than that which would be brought out by evidence.

2. by Acc. [Baldwin] Have you formed or expressed any opinions concerning any matters that in any way relates to this case?

Ans. [Hughes] I have formed but expressed no opinion.

14

2. by Acc. [Baldwin] Are these opinions which you have formed detrimental or prejudicial to the accused?

Ans. [Hughes] In case of Consideration, they are the same as the opinions discussed I would have against any duly enlisted soldier convicted of offenses for which the accused is supposed to be arraigned.

2. by Acc. [Baldwin] Have you at any time believed the accused to be guilty of the acts alleged against the accused?

Page 11 (3666)
Ans. [Hughes] I have.

2. by Acc. [Baldwin] Have you at any time considered on your mind the subject of the guilt of innocence of the accused?

Ans. [Hughes] Yes.

2. by Acc. [Baldwin] Have you talked about this case and do you know what the accused is charged with?

Ans. [Hughes] I have talked about it and I know what he is charged with.

2. by Acc. [Baldwin] Do you consider that having believed the accused guilty and with a knowledge of the offenses alleged against him that your mind is entirely free of all feeling and prejudice?

Ans. [Hughes] As the accused is one among a number [of] whom I believe guilty of the offenses with which he is charged my mind is not free of bias or prejudice.

Cross Examination

2. by J.A.[Judge] Would this bias or prejudice prevent you from doing justice to the accused and rendering a finding or verdict entirely in accordance with the evidence which may be adduced in this case?

Ans. [Hughes] No.

2. by Acc. [Baldwin] Please explain how being biased and prejudiced against the accused that you can direct your mind thereof and under judgment impartially and decide upon a question of guilt or innocence when that fact may only be determined by resolving a doubt on your mind for or against the accused.

Ans. [Hughes] I can form an opinion solely from evidence brought out in this case without reference to my previous belief or knowledge.

The Court was then cleared and closed, the alleged member retiring and after due deliberation the Court was re-opened. The accused and challenged member being present and the decision of the Court was announced by the Judge Advocate that the

Page 12 (3667)
Challenge is not sustained.

The accused then object to Lieut. W.G. Elliot, 9th Infantry, on the following grounds:

On the ground that he has formed and publicly expressed an opinion concerning the alleged acts of the accused and relating to his guilt or innocence. He is prejudiced against and on account of bias is not an impartial juror.

At the request of Counsel for accused the challenged member, Lieut. W.G. Elliot, 9th Inf., was there duly sworn by the Judge Advocate on his voire dire.

2. by Acc. [Baldwin] Have you any bias or prejudice for or against this defendant?

Ans. [Elliot] No.

2. by Acc. [Baldwin] Have you formed or expressed any opinions concerning the accused and relating to certain alleged acts which are supposed to have occurred on or about June 1st, 1887?

Ans. [Elliot] Yes.

2. by Acc. [Baldwin] Were those opinions prejudicial to him?

Ans. [Elliot] Yes.

2. by Acc. [Baldwin] Previous to being detailed on this Court, did you not believe that the accused was guilty of the acts alleged against him and occurring on June 1, 1887? And was not your opinion formed and expressed concerning him when that supposition?

Ans. [Elliot] Absent at the time of the occurrence and heard that a glaring outrage had been committed in the night at the community, I asked questions as to who the participants were and being told that it was "the Kid" and others I denounced them, on the evidence of witnesses. Since then I have talked about it as a matter of public notoriety.

Page 13 (3668)

2. by Acc. [Baldwin] Have you had at any time after the alleged occurrence of June 1, 1887 any bias or prejudice against the accused and has your conversations indicated bias or prejudice?

Ans. [Elliot] Yes. As biased against a man who has committed by common report a crime.

2. by Acc. [Baldwin] Who told you who the participants were?

Ans. [Elliot] Telegram partly from Capt. Pierce others Capt. Pierce by word of mouth others I don't know.

2. by Acc. [Baldwin] Were they officers?

Ans. [Elliot] I do not know.

2. by Acc. [Baldwin] Have you heard any officers at San Carlos express themselves concerning the accused in unfriendly terms and indicating bias and prejudice if so who were they?

 [Member] This was objected to by a member on the grounds of not affecting his competency as a witness.

[Judge] The Court was then closed for deliberations, the challenged member, Lieut. W.G. Elliot, 9th Inf., retiring and after due deliberations the Court was reopened and the Judge Advocate announced that the objection is sustained and the question will not be put.

The Counsel for the accused then asked the Court to have the following placed upon the record:

[Baldwin] I wish to have special upon the record the fact that Lieut. Elliott during the secret deliberations of the Court as to whether he, the said Lieut. Elliot should answer or not a question proposed to him, he was absent from the courtroom when the court was deciding upon the competency of the question.

"Under the law, custom of service and procedure of

Page 14 (3669)

Courts martial when a member is objected to, he retires, but when a member is a witness for any purpose even though it related to his competency as a member, he

17

must be present and take part in the deliberations at the court upon the relevancy and competency of the question. The court has no power to say beyond the law, who shall or who shall not take part in its proceedings as a member. Lieut. Elliot is a member at the present time and it is mandatory under the law until he be excused from taking part in this trial. That makes in all questions other than his competency. The objection relates solely to the competency of the question and not to that of the member and to deprive the accused of a dual role on any question affecting his rights and interests is illegal."

Motion[?] by a member, that the Judge Advocate be directed to spread upon the record that the absent member referred to was Lt. W.G. Elliot who being at the time under oath answering questions touching the validity of his competency as a member that the question objected to was put to this challenged member; that he was then not in position to deliberate in any of the proceedings during the closing of the court and for these reasons he was allowed to retire together with the prisoner and his Counsel. Su Par. Reg. 93—Ins. Military Law.

The Counsel for accused [Baldwin] objected to a member making a motion in the form of a reply to the counsel's remarks but to the member assuming the role of prosecutor.

The Judge Advocate replies that in his opinion this motion of a member could not make him appear in the light of a

Page 15 (3670)

prosecutor, but was simply an opinion of a member interview for the purpose of conferring the action of the court in excluding Lieut. Elliot. The Court had not yet had fully organized and Lieut. Elliot was a challenged member and it would have that the court would have the same right to excuse him at this time as to excuse him when the vote on the challenge as to his competency to serve as a member is taken. It is well known that a challenged member almost in_____ally relies during the deliberations for sustaining or not sustaining the challenge for the same reasons it deems proper that he should retire during all secret decisions affecting his competency to serve as a member when challenged. This question was one of importance and if it was necessary for him to vote on this it would seem that it would be necessary for him to vote to excuse or not to excuse himself as a member where challenged. See Inc. Military Law P. 93.

The counsel for accused [Baldwin] _____ to have it stated that after the remarks of the Counsel the Court was closed and that the Judge Advocate made his reply after the court was opened.

2. by Acc. [Baldwin] Have you not expressed yourself as to what punishment the accused should receive for the acts you believe he committed.

18

Ans. [Elliot] Not as an individual by name.

2. by Acc. [Baldwin] Do you consider that having believed the accused guilty and having had your mind once impregnated with prejudice by evidence of witnesses as to the occurrence as that you can try this case uninfluenced by bias or prejudice?

Ans. [Elliot] I do.

The court was then cleared and closed for deliberations the accused and the challenged member retiring and after due

Page 16 (3671)
Deliberation the court was reopened the accused and challenged member being present and the decision of the Court was announced by the Judge Advocate that the challenge is not sustained. [11]

The Court then at 5:40 o'clock P.M. adjourned to meet tomorrow at 8 o'clock A.M.[12]

L.D. Tyson [Lawrence Davis Tyson]
2nd Lieut. 9th Infantry
Judge Advocate

DAY 2
[HANDWRITING CHANGE, COURT REPORTER NO. 3 thin, whispy writing]
2nd Day of proceedings
San Carlos, A.T.
June 29th 1887

---

[11] The defense lawyer challenged all the members of the court for claim bias in Article 88 of the Articles of War at the time gave what they call a prisoner the right to challenge court members a court member could only be challenged one at a time and ultimately the members of the court would then vote on whether to excuse the court member challenge and in this case every court member was challenged and so the members decided by them forth for themselves whether they should be excused from the court-martial or not not surprisingly or the challenges to membership were all denied William Winthrop Colonel Winthrop and his treatise notes that there are a number of grounds for challenging members all of which seemed to be applicable here one is where a member forms an opinion or expresses an opinion about the outcome of a case and at least a couple of the members I believe 2d Lt. Hughes and 2d Lt. Elliot indicated that they had formed an opinion about the case which was adverse to the accused another was personal prejudice that they have a bias against the accused and it seems that virtually every member had some bias against the accused and finally being a material witness was a grounds for challenge and of course one of the members in this case Captain Bowman who was initially challenged for bias that was that challenge was denied ultimately moved to recuse himself because he was biased and as it turns out Bowman was a material witness in the case. [Rosen]

[12] Despite not being a lawyer, Lt. Baldwin's Voir Dire demonstrates they took their appointed jobs seriously. [Huffman]

The court met at 8:30 o'clock a.m.

Present:

1. Capt. A.H. Bowman, 9th Infantry
2. Capt. R.G. Smithers, 10th Cavalry
3. 1st Lt. H. DeLaney, 9th Infantry
4. 2d Lieut. C.P. Johnson, 10th Cavalry
5. 2nd Lieut. J.B. Hughes, 10th Cavalry
6. 2nd Lieut. W.G. Elliot, 9th Infantry

2nd Lieut. L.D. Tyson, 9th Infantry, Judge advocate.

The counsel and the accused also present.
[Judge]
The members of the court were there duly sworn by the Judge Advocate and the Judge Advocate was duly sworn by the President of the court; all of which oaths were administered in the presence of the accused.[13]
The Accused was then duly arraigned on the following
The court dispensed with the reading of the proceedings from the previous day.

Page 17 (3672)
Charges and specifications

Charge 1st
Mutiny 22nd Art. Of War
Specification 1st
[Who announces the charge?]
In that he, 1st Sergt. "Kid" Co. "A" Indian Scouts, a duly enlisted soldier in the service of the United States having been disarmed, and orderd to the guard house, by his Commanding Officer did disobey said order, and did in connection with others resist arrest seize his arms, open fire upon his Commanding Officer, and others connected with the military service and escape. This at San Carlos A.T. June 1st 1887.[14]

---

[13] The President of the Court as the senior member of the Court-Martial panel, served as Quasi-Judge ruling on evidence, etc. [Huffman]

[14] He was charged with mutiny including the fact that he "seized arms and opened fire on his commanding officer". In order to have a mutiny at the time, and I am looking at William Winthrop's treatise, mutiny was defined, and still is to an extent, as "an unlawful opposition or resistence to or defiance of superior military authority with a deliberate purpose to usurp, subvert, or override the same or to eject with authority from office." It is pretty clear from looking at the testimony in this case that the Kid did none of those things. There was no effort by the Kid to do any of these things. The only thing he did was to run away. The kid was present, surrendering his arms and that the shots were fired by the Scouts who were outside the tent on horseback. Nor is there is no evidence to indicate the kid seized

20

Charge 2nd
Desertion
In this that he, 1st Sargt. "Kid" Co. "A" Indian Scouts, a duly enlisted soldier in the .
service of the United States, did desert the same at San Carlos A.T. on the 1st day June
1887 and did remain absent, in desertion until he surrendered himself on the 25th day of
June 1887.[15]

Signed,
T.E. Pierce
Capt. 1st Infantry

[Baldwin]
The Counsel for the accused then applied to have plea of the accused made through
him which was allowed by the court. The accused then pleaded as follows:

To the specification 1st Charge 1st "Not Guilty"
To specification, 1st, Charge 2nd Plea in law of

Page 18 (3673)
Trial on the following grounds:
The law of trial: The essence of the offense of desertion consists in the intention not to
return to the service, or as it is expressed in orders from the War Dept. presumed to be
written on all ordered instituting courts the animus non revertende.[16]

---

any arms. In fact, the only evidence before the court was that he escaped and in fact he was unarmed
and he peaceably turned over his weapons and that he escaped without them.

So the most it seems to me that he could have been charged with was an escape as a prisoner which
was charged under a general article, which under the Articles of War at the time was Art. 62, today it is
Art. 134. It would have been a relatively minor offense. In fact, the only punishment for the escape of
prisoners at the time was Art. 69 of the Articles of War, which in fact punished officers who allowed
prisoners to escape.

[15] The desertion charge that was read indicated the Kid was absent without leave from a period beginning of June
to the 25th of June less than 30 days, when he surrendered himself to military authorities. Standing by itself, even
back in 1887, that was not sufficient to indicate the kid had deserted. All it indicated was that he had gone absent
without leave which was, and still is, a relatively minor offense. In order to establish desertion in 1887, the
government would have to prove that not only did the kid go absent without leave but he went with the intent not
to return to military authority. There was absolutely no evidence of that introduced during the trial. In fact, very
little was said about the desertion charge at all. He essentially escaped from arrest and disappeared for several
weeks.

Today, the sentence would be six months confinement and loss of two-thirds pay for six months, essentially a
misdemeanor. [Rosen]

[16] In the notion of a directed verdict of Not Guilty. [Huffman]

The specification upon its face shows that the accused voluntarily returned and surrendered and therefore negatives any presumption of desertion.[17]

[Judge] The court was then cleared and closed for deliberation. After due deliberation the Court was again opened, the accused and counsel present and the Judge Advocate announced that the plea in law of trial had been sustained by the court.

[Member] A motion to reconsider the action of the court in sustaining plea in law of trial was seconded [Member].

[Judge] The court was then cleared and closed.

[Judge]
After due deliberation the court was again opened and the Judge Advocate announced that the motion to reconsider had been sustained. The Judge Advocate then announced to the court that he would like to introduce evidence before the court to show that the plea in law of trial should not be sustained.

[Baldwin]
The counsel for accused objected to the Judge Advocate introducing testimony at this stage of the proceedings and requested that he be given until One o'clock in order to examine the law on the subject –matter before the court.
[Member]
A member of the court then moved that the court be closed to decide whether the plea in law of trial shall or shall not be sustained.

[Judge]
The court was then

Page 19 (3674)
cleared and closed.

After due deliberation the court was again opened and the Judge Advocate announced that on reconsideration the Court had decided not to sustain the plea in law of trial.

[Baldwin]
The counsel for accused asked for a short a time as possible in which to ascertain the law on the action of the Court and present are argument-why the Court should not reconsider the plea in law of trial already sustained.[18]

---

[17]

[18] A procedural objection. [Huffman]

[Judge]
The court then, at 10 o'clock A.M. took a recess until the counsel for accused could prepare his argument.

[Judge]
The Court again met at 11:00 o'clock a.m. and resumed its proceedings. All the members, the Judge Advocate, the accused and counsel being present.

[Baldwin]
The counsel for the accused there submitted the following argument:

May it please the court : The prisoner has plead to all charges and specifications. That plea has been accepted. This plea was in effect that the specification did not sustain the charge. It was a plea effecting the substance of the charge and specification. And in no way opened up the facts connected with the alleged offense other than is set forth as a legal indictment in the specification. The question is one of law and not of evidence. That the prisoner plead previous trial and acquitted, that could be property shown by evidence, but his plea, ~~settles all question~~, [written over this was "strikes on the question"] of offenses, made a necessary part of the offense by the law. The Court was determined whether a surrender a negatived on its face a non intention to return. The court at once

Page 20 (3675)
As decided, but upon reflection by a member's motion to reconsider the action of the Court was moved and seconded ---the court closed, after which the Judge Advocate makes a statement –writing to having evidence on the subject facts set out in the specifications and desires to introduce evidence. Facts and circumstances which are properly matters of Evidence are not legitimate subjects of pleas or in connection therewith. The Judge Advocate informed the court in effect, that he could or intended to establish the guilt of the accused, before the accused had admitted or denied that guilt and especially after the court decided the accused could not be tried upon the charge and specification. Th counsel had not time to present an argument. These are the circumstances still fresh in your mind. Now the accused has been arraigned and plead all of that has been accepted by the Court—Nay more, its decision was announced to the prisoner, that he could not be tried on the 2nd charge and specification at trial was begun I claim, as law and common justice shields the indian as well as the white man, that the court has no power to reverse its decision on a pleas accepted and formally announced. The court can consider its voet on a finding and sentence but not after it has been approved or left its control The plea of the prisoner had left your control when you accepted it and announced your decision and your jurisdiction only extended to determining the guilt or innocence of the accused on the matter before the Court. You confided your decision to the Judge

23

Page 21 (3676)

Advocate, he proclaimed it, the prisoner accepted it and stood before you duly arraigned your proceedings were complete and final as far as you have gone, to require the prisoner to plead again, where he has already pelad to every charge and specification – and you have accepted that plea, is to put him in perpetual jeopardy. Let me ask this Court one question and I have finished. Suppose the member who was dissatisfied with the Courts first decision and Captain Pierce, the witness, whow as present before the court waiting to be sworn, had been sworn and proceeded to testify and have shown by his testimony that the accused had no intentions to return to the service, or had other witnesses shown that fact at any subsequent period of the proceedings would that member have moved to reconsider the publically proclaimed decision of the Court: Could the Court then have granted it as it has done! If not, then why, in the name of humanity, which guards with sacred care the priceless interests of the humblest being charged with crime, do so at the time the Court did. Point of time is no element in this principle, under the circumstances and facts. The accused has been arraigned and plead; you have accepted that arraignment and plea and you have no power unless the law is trampled under to ask this defendant to plead again. The plea in law of trial, made by the prisoner says, the working of the specification denies that accused committed the offense charged against him. The facts set foth do not support the charge You have proclaimed that plea sound and valid. I ask that you wipe from the record your last action mollifying your previous action, which last action invades the rights of the accused.

Page 22 (3677)

The Court then proceeded with the arraignment of the prisoner and he then pleaded as follows:
[Baldwin]
To the 1st specification, Charge 2. Under the last decision of the Court, "Not Guilty" to the 2nd charge,
Under the last decision of the Court,
"Not guilty".
Captain A.H. Bowman then desired to submit the following statement to the Court.

[Bowman]
Referring to my reply of yesterday on being challenged on grounds of bias – after several hours have passed and during the arraignment of the prisoner, I deem it proper to say, before the Court proceeds further, that on mature reflection, I have grave doubts that I could vote on the findings in this case giving the prisoners the benefit of any doubt that might come up in mymind. In other words I find it difficult to divest myself of a decided bias against the accused before this court and those associated with him in the

24

outbreak at this place on the 1st inst These remarks are made in amendment to my remarks of yesterday.

On the statement submitted at this stage of the proceedings by Captain Bowman, the counsel for the accused against objected to Captain Bowman on the grounds of bias and prejudice against the accused.

Page 23 (3679)
[Judge]

The Court was cleared and closed at the request of a member.

The court was then opened again the counsel and accused resuming their seats.

[Judge]

The Judge Advocate then announced that the next business before the court was to act on the objection to Captain Bowman. The Court was then cleared and closed, the challenged member retiring.

After mature deliberation the court was reopened and the accused and his counsel were informed that no action would be taken on the objection made until communication and a reply thereto could be had from the convening authority.

[Baldwin]

The counsel for accused then submitted the following objection: When a court has jurisdiction has absolute jurisdiction over all matters submitted to its consideration and determination, the convening authority having the right to review its proceedings as a whole and approve or disapprove them, to permit the reviewing authority to pass judgement upon the statement of the members and the legal objection of the counsel, is to obtain his opinions upon the proceedings of this Court. The convening uahtority can dissolve this Court, but he cannot be asked to step in and review a part of the proceedings of this court until the entire proceedings, up to the present time have been submitted to him. To submit to the reviewing authority the questions whether or not this member is not competent to set upon this court, is to ask the reviewing authority, what action the Court

Page 24 (3680)
Shall take on the question before the Court. I object to anyone reviewing these proceedings at this time and advising the Court emmebers he appeared before the court, so that the counsel may be conversant with his proceedings. The members have an undoubted right and unquestioned right to amend his testimony on voir dire and the objection to his competency must be acted upon by the Court. The accused has that right to deprive him of it leaves him without protection of law."[19]

---

[19] This certainly would be a correct objection under current law. [Huffman]

The court was then cleared and closed. The court was then opened the accused and counsel resuming their seats and the Judge Advocate announced that he objection of the counsel had not been sustained. The court while in secret session[20] directed the Judge Advocate to present the salient facts relating to the testimony of Captain A.H. Bowman, 9th Infantry,to the Department Commander – to others facts relating to the status of the court . The Judge Advocate then after the adjournment of the Court sent the following message:

San Carlos A.T., June 19th 1887

To the Commanding Officer
Department of Arizona
Los Angeles, Cal.

Sir: I am directed by court to telegraph you that Captain Bowman, 9th Infantry, being challenged by counsel for accused yesterday on first meeting of court, placed under oath, stated that he had no bias nor prejudice against Seargt. "Kid" then on trial

Page 25 (3681)
Challenge not sustained. Today he made amendment to his statement of yesterday saying to the court that after consideration he finds it difficult to divest himself of a decided bias against the accused (Sgt. Kid) and those associated with him in the outbreak here June 1st. The court further says that "in view of the complications that have arisen, it is the opinion of the court that it adjourn until the facts are fully reported by telegraph, by the Judge Advocate to the convening authority – also that it is the sense of the court, that in view of the many complications of serious questions that are continually arising that it would be best that the court be dissolved and appoint ed of numbers who are not cognizant of the facts fo which the prisoners cited before it are being tried. Also that many members are liable to be called as witnesses and have been notified to that effect. The court is now adjourned waiting an Answer.
　　Tyson
　　Leiut. 9th Inft., J.A.

The court adjourned at 1 P.M. to convene again at the call of the President, After an answer had been received from the Department Commander.

DAY 3

---

[20] Based on the content of the letter written after the "secret session," I believe the members meeting in secret with the Judge Advocate were aware of the appearance of bias on the record. In that meeting they likely expressed the fact that they were all too close to the case, and maybe he could use this to move the case elsewhere by writing this letter to the convening authority. [Sutton]

[HANDWRITING CHANGE, COURT REPORTER NO. 4]
L.D. Tyson
2nd Lieut., 9th Inf., Judge Advocate
San Carlos, A.T.
July 2, 1887
[Judge]
The Court met pursuant to adjournment at 8:15 o'clock A.M.
The order convening the Court was then read.

Page 26 (3682)
The following telegraphic order was then read from the official telegram received by the Judge Advocate:

[Brig. Gen. Miles]
War Dept. Signal Service U.S.A. Telegram
Los Angeles, Cal
June 30, 1887

Lieut. L.D. Tyson, 9th Infantry: Judge Advocate. General Court Martial

The Following order was issued today. Paragraph one. S.O. sixty-seven. Major Anson Mills, First Lieut. L.A. Hunt and R.D. Read Jr., 10th Calvary, are detailed as members of the General Court Martial convened at San Carlos by Field Orders One C.S. upon completion of the cases before the court they will return to their proper stations. The journeys as directed ar necessary for the public service.
By command of Brig. Genl Miles
(Sgt. McBarber Asst Adjt General)

The court met pursuant to the foregoing order at 8:15 A.M. Present

1. Major Anson Mills, 10th Calvary
2. Captain A.H. Bowman, 9th infantry
3. Captain R.G. Smither, 10th calvary
4. 1st Lieut. L.P. Hunt, 10th calvary
5. 1st Lieut. H. DeLang, 9th Infantry
6. 1st Lieut. R.D. Read, 10th calvary
7. 2nd Lieut., C.P. Johnson, 10 calvary
8. 2nd Lieut., J.B. Hughes, 10 calvary

Page 27 (3683)
   9.   2nd Lieut. W.J. Elliot, 9th Infantry
2nd Lieut. L.D. Tyson, 9th infantry, Judge Advocate

The accused and counsel also present.

The following objections was then made by counsel for accused.

[Baldwin]

I object to Major Mills Lieuts. Hunt and Read, taking part in this case, for three reasons, under the order just read. There is no precedent for members to ask upon a telegraphic order detailing them upon a court Martial. The order is not in proper form, and is only official information to Lieut. Tyson Judge Advocate, that the order (on to use the words of the telegram, the following order was issued today) was issued. To ask a Court must have the official order on the accused has the right to travense its legality at present there is no legal evidence that, that telegram is genuine or that said purported order was actually issued, by the proper authority.

[Member]

The court was then on motion of a member cleared and closed and after deliberation it was decided that a proper mode of procedure would be that the original Court should act on the previous objection of the accused to Captain Bowman as in the regular order of business before any participation of the new members in the business before the court at present. The court was then re-opened. The counsel and accused resuming their seats. The Judge Advocate then announced the decision of the Court. The new members then retired.

[Member]

At request of a member of the Court the Judge Advocate then read the amended testimony of Captain Bowman,

Page 28 (3684)

Together with objection of counsel for accused to Captain Bowman's remaining on the Court. The Court was then cleared and closed. The challenged member retiring.

After due deliberation of the Court was then reopened challned member and accused resuming their seats and the Judge Advocate announced that the challenge has been sustained and Captain Bowman will be excused from sitting as a member in this case. Captain Bowman retired.

The new members of the Court then resumed their seats. The objection of the counsel read sitting on the court , was then read to the Court. The Court was then cleared and closed.

[Judge]

After due deliberation the Court was reopened the accused and Counsel resuming their seats and the Judge Advocate announced that the objection has not been sustained.

The Counsel for accused then commenced the reading of a paper which is attached to these proceedings and marked "b',[Exhibit B is 9 pages long and is an impassioned plea not to impanel these members or it will be a miscarriage of justice.] as for as the word Major Mills. The Court was then ordered cleared and closed. After due deliberation the court was again opened and the counsel for the accused was informed that the court had decided that he must confine himself to the matter at issue before the court, and to make objections as brief as possible to convey their purpose.

The presiding Officer informed the Counsel that all communications made by him to the court must first

Page 29 (3685)
be passed to the Judge Advocate—

[Baldwin]
The counsel for accused then made the following objection: The Court having decided than a order was before the Court upon which it could act, I now object to order itself as illegal and in support of which desire to rpesent an argument to the Court. (This paper is attached and marked "B") The Counsel for accused then stated to the Court that he did not desire that the argument should be interpreted to the accused. The JDuge Advocate replied by quoting from Winthrops Digest of opinions of Judge Advocate General, Pargraph 3, page 320. The Counsel invited the attention of the Court to G.O. No. 3 1881 War Department.

The court was then cleared and closed. After due deliberations the Court was then reopened the accused and counsel also present and the Judge Advocate announced that the objection has not been sustained. The Judge Advocate then asked the Counsel for accused if he had any objection to any individual member of the Court as now organized.

[Baldwin]
The Counsel then desired to have Major Mills put on his voire dire. Major Mills was then duly sworn on his voire dire by the Judge Advocate and testified as follows:

2. by Acc. [Baldwin] Have you any bias or prejudice for or against this defendant or formed or expressed an opinions concerning this case.

Ans. [Mills] No sir; I do not know what he is charged with.

Page 30 (3686)
[Baldwin]
The Counsel then stated that he had no objection to Major Mills.

[Baldwin]
He then asked that Lieut. Hunt be placed on his voire dire as foundation for an objection: Lieut. Hunt was then duly sworn on his voire dire by the Judge Advocate and testified as follows:

Questions by accused.
[Baldwin]
Have you any bias or prejudice for or against this defendant; or formed or expressed any opinions concerning this case?

[Hunt]
I have not.

[Baldwin]
The Counsel then stated that he had no objections Counsel then asked that Lieut. Read be put on his voire dire as foundation for an objection. Lieut. Read was tehn duly sworn on his voire dire by the Judge Advocate and testified as follows:

[Baldwin]
Have you any bias or prejudice for or agasint this defendant. Have yo formed or expressed anyopinions concerning this case, so far as it related sot the supposed conduct of the defendant "Kid".

[Read]
I have no bias or prejudice against him nor have formed nor expressed any opinions as to his particular conduct. I do not know what the prisoner is charged with.

Question by accused
[Baldwin]
Have you heard of the occurences supposed to have taken place at this post Early in last month and relating to an outbreak of Indians, if so, have you formed

Page 31 (3687)
Or expressed an opinion concerning the supposed acts of those persons.

Answer:
[Read]
I heard of an outbreak occurring here some time last month. They were rumors from persons who were not present at San Carlos at the time of the alleged outbreak. These were conflicting rumors and I did not form any definite opinion as to who had been involved in it or as to the guilt or innocence of any body.

Question by accused.
[Baldwin]
Did you not hear that Captain Pierce and others had been fired upon by Scouts if so did yo not believe that report.

Answer:
[Read]
I heard that captain Pierce had been fired on by a party of Indians and that Scouts were in the party but I did not hear that scouts did any firing.

Question by accused
[Baldwin]
Did you believe the report that Captain Pierce and others had been fired upon by a party of Indians.

Answer
[Read]
I did not know which one to believe.

Question by accused:
[Baldwin]
You have said you had formed no definite opinion, now did you form any opinions that were prejudicial to those whom you had heard committed the acts mentioned by you?

Answer:
[Read]
I had not heard that nay particular persons committed such acts. I merely formed a vague idea from the stories that I had heard that this had been some disturbance down here but I formed no prejudice against any particular

Page 32 (3688)
persons.
[Baldwin]
The counsel for accused then made the following objection.
I object to Lieut. Read on the ground of prejudice, bias and malice agasint this defendant. The member then stated that he had no bias or prejudice against the accused that might do him an injustice as a member of this court.

[Judge]
The court was then cleared and closed the challenged member returning. After due deliberation the Court was again opened the accused and challenged member resuming their seats and the Judge Advocate announced that the challenge is not sustained.

[Baldwin]
The counsel had no further objection to any member.

[Mills?]
The President of the Court then directed the Judge Advocate to swear the whole Court.

[Baldwin]
Counsel made the following objection: I object to the proceedings. The swearing of this whole Court. This court has been sworn, or at least a legal number of members have already been sworn as a court including the Judge Advocate.

[Judge]
The Court was then cleared and closed. After due deliberation the court was again opened, the accused and counsel resuming their seats, the Judge Advocate announced that the objection had not been sustained.
All the members of the Court were then severally duly sworn by the Judge Advocate and the Judge

Page 33 (3689)
Advocate was duly sworn by the President of Court, all of which oaths were duly administered in the presence of the accused.
The record of the proceedings of the previous days was then read by the Judge Advocate to the court. The Court then at 12 noon took a recess until 1:30 o'clock p.m. The court at 1:35 o'clock p.m. resumed proceedings all the members, the Judge Advocate, the accused and counsel being present.

[Judge]
It was then announced to the counsel that he was offered an opportunity to withdraw the plea of the accused and renew the plea in law at trial for the accused if he so desired.

[Baldwin]
Counsel for accused replied:
The accused has plead to all accusations against him, he has nothing to say now on the question of please. He has been arraigned before this court and has plead and is now on trial. Hence there is nothing for him to do but demand proof of the accusations set forth against him. There is a telegram before this court being as I understand a reply from the Department Commander or reviewing authority to the communication from the court. I desire that it be pleaced upon the record. Many of the Members have individually been shown that telegram. I have seen it. The President of the Court has announced that he decides. If there is no objection that it is not necessary to place that telegram upon the record as not relating to the case. I deny his right.

Page 34 (3690)
I desire to have the opinion of the court on the question and I further ask as that telegram is ananswer to the Communication from this Court to the Department and as

that Communication is now a part of the record, that the answer be also recorded, as a part of the proceedings of this case. The matter having come up in open court.

The President of the Court stated that unless there was some objection by some member of the Court to the Judge Advocate the telegram of the Department Commander would not be considered a part of the proceedings. As no objection was made by any member the telegram in question was not made a part of the record.

There being objection the court was cleared and closed. After due deliberation the Court was reopened the accused and counsel resuming their seats and the Judge Advocate announced that the court had decided to put the telegram in question on the record as follows:

War Department, Signal Service, U.S. Army (Telegram)
Los Angeles, Cal.
June 30, 1887

Lieut. Tyson, 9th Infantry, Judge Advocate, Genl. C.M.
                    San Carlos, A.T.

The Dept. Comdr sees not legal objections why the court as organized should not proceed with the trial of cases of Indians. It still has a legal number of members and organization. We have however ordered from Grant an Thomas as additional members of the

Page 35 (3691)
Court. Major Mills and Lieut. Hunt and Read
        (sgd) Mr. Barber
        Asst Adjt General

[Judge]
Captain T.E. Pierce a witness for the prosecution was then duly sworn by the Judge Advocate and testified as follows:

[Judge]
Question by Judge Advocate-
State your name, rank and Station.

[Pierce]
Answer:
T.E. Pierce, Captain 1st Infantry, San Carlos, Arizona

Question by Judge Advocate-

Do you know the accused and if so ~~by what name~~ as whom?

[Pierce]
Answer: I do. Late 1st Ssergeant Kid, Company "A" Indian Scouts.

Question by Judge Advocate:
How long have you known him?

[Pierce]
Answer: About two years.

Question by Judge Advocate.
What was the occupation or calling of the accused on June 1st 1887.

[Pierce]
Answer: He was 1st Sergeant Company "A" Indian Scourts in the Military Service of the United States

Question by Judge Advocate.
Was he a duly enlisted Sodier?

[Pierce]
Answer: He was. He was enlisted by Capt. P.L. Lee, 10th Calvary on the 11th day of April 1887 at San Carlos A.T.

Question by J.A.:
[Judge]
Do you know if the oath was duly interpreted to him and by whom.

[Pierce]
Answer: I do. I don't remember if it was by McIntosh or Antonio Diaz. I think by Antonio Diaz.

Question by Judge Advocate:
Did you see the accused on June 1st 1887?
Page 36 (3692)

[Baldwin]
Answer: The accused Counsel for accused objected to this question on these grounds: The question is leading and incorporates a material allegation into the question, by informing the witness the particular day upon which the occurrence is suppoed to have taken place.

The Court was the cleared and closed. After due deliberation the Court was reopened the accused and Counsel resuming their seats and the Judge Advocate announced that the objection had not been sustained

[Pierce]
Answer: I did.

Question by Judge Advocate:
Where and under what circumstances did you see him?

[Pierce]
Answer: I saw him about sundown in front of Mr. Sieber's tents. He absented himself from camp without permission on the evening of the 29th or morning of the 28th of May 1887. On the 30th of May the messenger alleging that he came from Sargeant Kid, said that he "Kid" wished to see me, I replied that he could see me if he wished to but that I should make no promises whatever. On the morning of the 1st of June the messenger "gonshayee" came again and said that he came from Kid and that "Kid" wished to see me. I told him that he could come if he pleased and that the sooner he came the better it would be for him. About sundown on the first of June I was informed that he had returned and that he was at Mr. Seiber's quarters. I went ther with Mr. Seiber and also with Antonio Diaz

Page 37 (3693)
And Fred Knipple as interpreters. Four other Scouts who absented themselves at the same time that "Kid" did were at Mr. Seibers tent with him. When I reached there I said, where are the five scouts who have been absent. They all stepped to the front in line I said to "Kid" give me your rifle. He handed it to me and I said give me your belt. He handed it to me and I put it on a chair that was standing there. The same Conversation and occurrences took place with the other four Scouts. I then said "Calaboose." Two or three stepped to the chair picked up their belts and commenced removing their knife scabbords and knives from their belts—they being their person property. I also took hold of a belt and commenced taking a knife scabboard off. Just then I heard a little noise that attracked my attention in front of the tent and I saw a few men on horseback who were bringing down their firearms and getting cartridges from their belts. I said lookout Sieber they are going to fire. Immediately there was a shot fired by some one in the party in front of the tent. I turned round walked into the front door of the tent and and through the tent. The tent had two doors , one front and one rear. Both open. Sieber got down and crawled into his tent. Also Knipple. I did not see what became of Antonio Diaz. There were about fifteen to thwety-five shots fired by persons in front of the tent, quite a number of them passing through the back of the tent, all entering the front door. I came back through the back door of the tent and through the front door.

Page 38 (3694)

Sieber was in his tent wounded though the left leg below the knee. The belts that I had left in the chair had all disappeared and also two of the fire arms, that I had taken from the Scouts. Those rifles were lying on the ground near where I had stood them after I had taken them from the Scouts. I picked them up carried them into Mr. Siebers tent and put them on his bed. The five Scouts and some Indians who were in front of the tent besides then had disappeared and the firing had nearly ceased when I came to the front of the tent again. I can not identify any particular one as having fired a shot. After they disappeared form the front of the tent on the evening of the 1st of June they were not seen by me again until the 24th of June.

Question by Judge Advocate:
In what capacity were you acting when you ordered the accused to the Calaboose.

[Pierce]
Answer: As being in Command of a post and as in Charge of police control of the White Mountain Indian Reservation.

Question by Judge Advocate:
In whose charge did you consider this party of five Scouts of whom the accused is on after you gave the order "Calaboose".

[Pierce]
Answer: Under change of the Guard.

Question by Judge Advocate:
What authority did the accused have for absenting himself from his post and duty?

[Pierce]
Answer: He had no authority or permission.

Q by J.A.
[Judge]
Do you know what connection or relation the accused had with or to the parties who were present at Mr. Siebers tent at the time in question?

[Baldwin]
The Counsel for the accused objected on following grounds:
The question is leading, irrelevant, immaterial and incompetent. The accused is not charged with a conspiracy. He is on trial for his individual acts, and the question must relate to the points at issue.

Page 39 (3695)

36

[CHANGE IN HANDWRITING, Court Reporter no. 5. Uses capital cursive Es everywhere, and misspells "cheif" repeatedly and other words.

[Judge]
The Judge Advocate replied that the accused is charged with mutiny which is defined to be in general a concerted procedure; the concert itself going for to establish the crime. Besides specifications directly allege that he did disobey orders and in connection with others resist arrest.

The court was then cleared and closed, after deliberation the court was opened the accused and counsel resuming their seats and the Judge Advocate announced that the objection had not been sustained.

[Pierce]
Answer:
All that I saw there to recognize fully with two or three exceptions belonged to the same band that Kid does San Carlos Indians."I"

2 by Judge Advocate
What was the general bearing and appearance of the accused. The four scouts and those on horse back during the time that the scouts were being disarmed. That is: was it hostile or friendly, or what impression did it make onyou at the time?

[Pierce]
Answer:
The five scouts appeared to be perfectly obedient and perfectly willing to obey the orders I gave them. I expected them to start directly for

Page 40 (3696)
the Calaboose. I did not notice the bearing of the others until my attention was attracted by an unusual commotion, there they were bringing down their arms and getting cartridges and their horses were restless.

2 by Judge Advocate-What has been the general relation of the accused to his band and what is his general standing in the band?

[Baldwin]
The counsel for the accused made the following objection:
Irrelevant and immaterial, the court is trying the accused and not any band of Indians and his relations to other Indians not Soldiers or his standing or bearing in a band of Indians unconnected with the service has no bearing upon the case. If the question is permitted the Judge Advocate may ask next his relations bearing and standing among white people. The Commanding Officer has no connection with the Indians at this post,

other than Scouts, and therefore the question relates more particularly to the knowledge and duties of an Indian

Page 41 (3697)
Agent.

[Judge]
The Judge Advocate states: I desire to show by this question that the accused was such an influential man in his band that he had induced these parties to come to San Carlos with him armed to do his bidding and prevent the authorities here from taking him to the Guard House. Also to show that he was such an influential man that he had once been named for Chief but would not accept and that he had such influence that he could work his will on the others.

The court was then cleared and closed. After due deliberation the court was again opened the counsel and accused resuming their seats and Judge Advocate announced that the objection had not been sustained.

[Pierce]
Answer:
His Grandfather was cheif [sic] of the band. The cheif [sic] died last December. Another man was selected by the band as cheif [sic] in his place. So far as I know the accused is held in high esteem by the members of his band and particular since the death of his grandfather.

Page 42 (3698)
2 by Judge Advocate.
Has he ever been elected or offered any office in the band and if so state what was that office.

[Baldwin]
Counsel for accused then made the following objection. Irrelevant and immaterial. The questions must relate to the point at issue. This evidence must have a bearing on his acts as a soldier of the United States Army and he is not charged with having had a revolt, or engage in war against the United States and as to whether he is or has been elected a chief of his band has no bearing on this case before the court. The testimony of the witness already shows all facts, so far as he is concerned connected with the alleged mutiny.
[Judge]
The Judge Advocate replied that he had put this question for the same reason that he had put the one just before it. The counsel then replied as follows: The past history and character of the accused is not now before the court.

The court was then cleared and closed after due deliberation the court was reopened the accused and counsel resuming their seats and the Judge

Page 43 (3699)
Advocate announced that the objection was not sustained.

[Pierce]
Answer:
He has never been elected to any office that I know of. After the death of his grandfather, when they were consulting as to who should be chief in place of his grandfather my recollection is that his name was mentioned and successor to his grandfather, but they finally selected a man named Gow-shay-ee.

2 by Judge Advocate
How many days had the accused been absent without authority?

[Pierce]
Ans:
Thirty days. Five days the first time, and twenty five days including the first of June afterwards.

CROSS EXAMINATION.

[Baldwin]
2 by accused.
Was he the popular choice for chief of his people or was his name only mentioned as a candidate for office.

[Pierce]
Answer:
He was not the popular choice or I suppose he would have been selected.

[Baldwin]
Question by accused
Has there been any discord or dissatisfaction existing among his people at his not having been made chief.

[Pierce]
Ans:
Not that I know of.

[Baldwin]

Question by accused
Has the accused any considerable

Page 44 (3700)
following among his band more than any other indian, who is generally liked by his people.

[Pierce]
Ans:
I think he has considerable influence in his band. Thirteen men of his band went with him, while six stayed at home and did not go.

[Baldwin]
Question by accused:
Did Gonshayee absent himself at this time, and among the number you have just mentioned.

[Judge]
The Judge Advocate objected to this question on the grounds that nothing had been said by the witness about Gon-shay-ee absenting himself

The Judge Advocate then read the testimony of the witness on the examination in chief to the court relating to Gonshawee.

[Baldwin]
The counsel for the accused then made the following objection reply:
The question of the accused having enticed by his prominence, some of his people to aid him in a supposed mutiny and therewith then fled has been raised by the Judge Advocate. He has submitted testimony on the point, even to the extent o showing who is chief of the accused's band. He has gone further

Page 45 (3701)
And endeavored to show, the prominence and influence of the accused among his people, and the witness has said that the influence of the accused is such that thirteen men left with him. The witness further has stated in direct examination as to about the number of Indians and scouts that absented themselves. Now I claim we have a right to know who these Indians were if the witness knows.
Further a cross examination is confined to all matters brought out on the direct examination, cross examination and upon what has been brought out by other witnesses.

The court was then cleared and closed. After due deliberation the court was reopened the accused and counsel resuming their seats and the Judge Advocate announced that the objection had not been sustained.

[Pierce]
Answer:
He did. I mean on the first of June. He came in on the twenty-second of june.

[Baldwin]
Question by Accused:
During May when the accused was absent as you have stated whom did you send to carry your message to him.

Page 46 (3702)
[Pierce]
Answer:
I did not send any out.

[Baldwin]
Question by Accused:
Did you see the accused take his arms or any arms, at the time of the occurences mentioned by you at Siebers tent.

[Pierce]
Answer:
I did not.

[Baldwin]
Question by accused:
State about the number of Indians at or about Sieber's tent at the time of firing. That you mentioned as having occurred.

[Pierce]
Answer:
I should think about twenty .

[Baldwin]
Question by accused:
Did the accussed cheerfully and obediently start to comply with your orders.

[Pierce]
Answer:
He handed me his rifle

Page 47 (3703)
And belt without saying a word as soon as I asked him for them.

[Baldwin]
Q by Acc:
When you said they are going to five whom did your refer to ?

[Pierce]
Ans:
To some people who were on horseback, men of his band about twelve feet in front of the tent.

[Baldwin]
2 by Acc:
When the first shot was fired from whence came it and in what direction was that shot fired.

[Pierce]
Ans:
It seemed to me to come from about the front of the center of the tent. IT seemed so to me. I could not exactly tell.

[Baldwin]
2 by Acc:
Where were you standing and how far from you was the shot fired.

[Pierce]
Ans:
I was standing just in front of the center of the tent. Just under the arbor. I suppose the twelve or fifteen feet from where the shot was fired.

[Baldwin]
2 by Acc:
At the time the shot was fired and at the time you said they are going to fire where was the Kid?

[Pierce]
Ans:
My recollection is that he was standing close to the chair that I had put the belts on and close to Sieber and myself. I am not certain

Page 48 (304)

of that---

[Baldwin]
2 by Acc:
Did Sieber fall and crawl as a result of that shot.

[Pierce]
Ans:
I don't think he did until after several shots were fired. The shots came almost all together after the first one.

[Baldwin]
2 by Acc:
Did you see him wounded

[Pierce]
Ans.
I did not.

[Baldwin]
2 by Acc:
Did the accused have time to obey your directions to go to the Calaboose. And did he not make preparations to do so.

[Pierce]
Ans.
He did not have time to go. He had time to start I supposed he was going he appeared to be making preparations.

[Baldwin]
2 by Acc:
At the time of the disturbance, riot, or revolt indicated by the firing you first mentioned did you call upon the accused as a soldier to suppress it.

[Pierce]
Ans:
I did not.

[Baldwin]
2 by Acc:
At the time of the conference at Siebers tent was any guard present in anticipation of trouble.

[Pierce]
Ans:
There was not. There was some scouts standing there intended to send one of them with the prisoners

Page 49 (3705)
to the Guard House. I had not ordered any scouts or soldiers to be there.

[Baldwin]
2 by Acc:
By what authority are you exercising your police control at this agency.

[Pierce]
Ans:
By orders of the Department Commander.

[Baldwin]
2 by Acc:
Are you not also acting in the capacity as an Indian Agent of the Interior Department.

[Pierce]
Ans.
I am.

[Baldwin]
2 by Acc;
Have you all the time during the past two years been Commanding Officer at this post.

[Pierce]
Ans:
I have not.

[Baldwin]
2 by Acc:
Does this accused in your opinion understand your dual Officer of Indian Agent and Commanding Officer.

[Pierce]
Ans:
I think he does.

[Baldwin]
2 by Acc:

As a soldier has the accused been taught to distinguish between the offenses against a Commanding or other Officer, and those against others not Officers.

[Pierce]
Ans.
He is sworn to obey the orders of the President of the United States and of Officers appointed over him and this is carefully interpreted. I think that he has about as much regard for one officer as another and that although

Page 50 (3706)
he understands who the Commanding Officer is he would think it no worse to commit an offense against the Commanding Officer than any other officer.

[Baldwin]
2 by Acc:
Has the accused received to your knowledge any instructions in the matters found in the rules an darticle of war.

[Pierce]
Ans.
The articles of war have never been read and translated to him to my knowledge. He has been verbally instructed in his duties as a soldier.

REDIRECT EXAMINATION

[Judge]
Q by J.A.:
You say that Kid had time to start to the calaboose after you ordered him there. Did he start at all so far as you know?

[Pierce]
Ans:
I do not think he did.

[Judge]
2 by JA:
Did you see him while he was going to Sieber's tent?

[Pierce]
Ans:
I did not.

[Member]

45

Q by Member
Is Gowshayee a man of strong personal influence among his people?

[Baldwin]
The counsel for accused objected to this question on following grounds:
I object to the question on the ground that it is not proper for a member there of to originate evidence

Page 51 (3707)
beyond clearing up doubts upon dubious answers. I further object upon the ground that the question is irrelevant and immaterial and not confined to the point at issue.

The court was then cleared and closed. After due deliberation the Court was then reopened the accused and counsel resuming seats and the Judge Advocate announced that the objection has not been sustained. The question was then recorded as a question by the court.

[Pierce]
I do not think he is. He has complained to me several times that he could not do anything with them. That they would not obey him.

[Court??]
Q by the Court
Are you satisfied that the Indians enlisted as scouts fully understand that disobeying an order of their Commanding Officer and firing upon him is a crime of the gravest character?

[Pierce]
Ans.
They understand that it is a very grave offense.

The Court was then at 6:30 P.M. adjourned to meet again at Nine o'clock A.M. tomorrow.

L.D.Tyson
2nd Lieut., 9th Infantry, Judge Advocate

DAY 4
[Handwriting changed back to Court reporter no. 4]
Page 52 (3708)

San Carlos A.T.

July 3, 1887

The Court met pursuant to adjournment at 9 o'clock a.m.

Present

Major Anson Mills, 10th Cavalry
2 Captain R.G. Smither, 10th Cavalry
5 1st Lieut. L.P. Hunt, 10 Cavalry
4 1st Lieut. H. DeLany, 9th Infantry
5 1st Lieut. R.D. Read, 10th Cavalry
6 2 Lieut. C.P. Johnson, 10 Cavalry
7 2 Lieut. J.B. Hughes, 10 Cavalry
8 2d Lieut. W.G., Elliot, 9 Infantry, and
2d Lieut. L.D. Tyson 9 Infantry, Judge Advocate

The accused and counsel also present.
The Court dispensed with reading the proceedings of the previous day.
The Court met at Sieber's Room on account of the witness being too sick to appear at the regular courtroom.

Then moved and seconded that the court adjourn until tomorrow at 8 o'clock.
The question was then put to the court and the motion was lost.

Al Sieber, Chief of Scounts, a witness for the prosecution appeared before the court, was then duly sworn by the Judge Advocate-and testified as follows:

[Judge]
Q by J.A.
State your name rank and station

[Sieber]
Ans.
Albert Sieber, occupation Chief of Scouts; and residence Agency of San Carlos, A.T.

[Judge]
Q by J.A.
[margin note reads: Interlineations made by direction of Judge Advocate, as if he went back and added things that were not said in this order or said at all. Then a line written above the question here reads,]
Do you know the accused and if so is whom?

[Sieber]

47

Ans:
He was Kid 1st Sergeant "A" Company Indian

Page 53 (3709)
Scouts.

[Judge]
Q by J.A.
How long have you known him?

[Sieber]
Ans.
About 8 years.

[Judge]
Q by J.A.
What was his occupation.

[Sieber]
Ans.
He was 1st Sargeant of Indian Scouts.

[Judge]
Q by J.A.
Where were you on June 1st 1887?

[Sieber]
Ans.
I was here at San Carlos Agency

[Judge]
Q by J.A.
Do you know if he is a duly enlisted Soldier of the United States.
[Sieber]
Ans.
I know that he is duly enlisted according to the law of the United States as a scout.

[Judge]
Q by J.A.
When, where and by whom enlisted.

[Sieber]
Ans.

He was enlisted here on the 11th of May 1887 by Captain Lee, and the oath was duly interpreted to him by Antonio Diaz and Fred Knipple. I think the time was the 11th of May.

[Judge]
Q by J.A.
Did anything unusual occur here on June 1st 1887. And if so state what it was and its cause.

[Sieber]
Ans.
On June 1st Kid the 1st Sargeant of Scouts who had been anbsent for five days without permission returned along in the evening about 5 o'clock I should judge, and I saw him coming up to my liet. I happened to be up here at the agency and went down and reported it to Captain Pierce that they were at my tent, the Captain then walked down with myself to my tent; on arriving there he asked me to point out the five Indians who had been absent without leave and returned. As I did so he disarmed them, took their guns from them, belts and after receiving their arms and belts he ordered them to the Calaboose. He gave the

Page 54 (3710)
Order two or three to the 1st Sargeant Kid and the other four Scouts, Kid gave a look which signified

[Baldwin]
The counsel for accused then made the following objection.
I object because the witness is testifying to his own conclusions and deductions. It is for the court to determine what the look of the accused signifced.

The Court was then cleared and closed, after due deliberation the Court was again opened the counsel and accused resuming their seats and the Judge Advocate announced that the objection had not been sustained by the Court.

[Sieber]
The witness then proceeded.
To me for them all to rush for their arms which they did all five of them. Captain Pierce jumped in between two of the scouts and their arms and shoved their arms away fom them, and I jumped between the three others and their guns. Kid being the nearest to me made a grab for his Carbine which I grabbed with my right hand and shoved Kid back with my left and about this time Captain Pierce hllared to me, "Look out Sieber they are going to shoot, and he broke for my tent, that is the Captain. I turned and kicked the guns that were in my reach towards my tent as far as I could At this time

49

there were two shots fired one right after the other, Kid jumped away from me and ran around the shade of my tent, and I then ran into my tent throwed myself down to get my gun. At this time there

Page 55 (3711)
More scattering shots being fired that I could hear going through my tent, some of them. I by this time got my ugn and went out doors and took a shot at the first Indian I saw who was on horseback, evidently had just fired, at the same time that I fired I was shot in the leg my leg broken and I was knocked down. I fell back into my tent and found Fred Knipple there and he commenced bringing water for my leg.

[Judge]
Q by J.A.
Did you see the accused coming to your tent in company with any one else before you reported the matter to Captain Pierce on June 1st and if so state who they were.

[Sieber]
Ans:
I have the list of names, but the list that I generally keep I have their tag numbers opposite their names. The witness then referred to a memorandum in order to get their names. There were "Kid" Osksalaha, Sergeant Company "A" Indian Scouts. Nacanquisay. Corporal Co. "A" Indian Scouts Bachoonhah, Private Company "A" Indian Scouts and Margy, Private Company "A" Indian Scouts.

[Judge]
Q by J.A.
Who were the other Indians who belong to San Carlos "I" band. Gonshayee Chief.

[Judge]
Q by J.A.
Did they belong to the band as the accused?

Page 56 (3712)
[Sieber]
Ans:
Yes, they belonged to the same band as Kid. Not all the five scouts belonged to this band, but four of them did.

[Judge]
Q by J.A.

50

What was the appearance and attitude of these Indians on horseback when you came up to your tent at the time stated by you?

[Sieber]
Ans
My opinion as to their appearance is that it was war-like, I thought so by their looks, and by them having arms in their hands which is against orders here,

[Judge]
Q by J.A.
What was the appearance and attitude of the five Scouts at this same time?

[Sieber]
Ans.
They looked cross as I came up and went by the. I spoke to Kid and he answerd me gruffly.

[Judge]
Q by J.A.
Did the accused start or make any attempt to go to the guard House when Captain Pierce orderd him to go?

[Sieber]
Ans.
Yes. I thought that Kid did nake a movement as though he wanted to go.

[Judge]
Q by J.A.
In what direction were the shots that you heard directed?

[Sieber]
Ans.
They were coming from in front of my tent like towards the tent, several of them went through the tent, some came from in front and some from the side.

[Judge]
Q by J.A.
Did any of those shots come close to Captain Pierce or in the direction of the place where he was standing at this time?

[Sieber]
Ans:

The first two shots that were fired must have come very close to Captain Pierce as I think they were fired right at him. A number of hsots moore went through the tent in the same direction that Capt. Pierce went.

[Judge]
Q by J.A.
Did any one else jump for his arms at the time you mentioned that Kid jumped for his.

[Sieber]
Ans.
All five of the Scouts jumped for their arms at

Page 57 (3713)
the same time.

[Judge]
Q by J.A.
Did this unanimity of action appear to be accidental.

The Counsel for accused objected on following grounds:

Page 58 (3714)
[Baldwin]
I object. For the reason that the question is incompetent. It is for the members of the court to determine concert of action, from all facts, and circumstances connected with the conduct of the accused. In crime there is intent and to show that intent facts must be resorted to and not the opinion of witnesses.

[Judge]
The Judge Advocate replied that he desired to show by this question and others that may follow, that there was preconcerted action on the part of the accused, and the other four scouts with him, tending to show the gist of the offense of mutiny: concerted action.

The court was then cleared and closed. After due deliberation the court was reopened the accused and the counsel present and the Judge Advocate announced that the objection is not sustained.

The accused then answered!
[Kid]
No.

[Judge]

52

Q by J.A.
How did you interpret the look that you mention as having been given to the other four scouts by the accused at the time they all attempted to get possession of their arms.

[Sieber]
Ans.
I interpreted it that Kid gave the look to these four scouts, after the captain telling them twice to go to the calaboose. I thought they had an understanding

Page 59 (3715)
Before they came in here, that if a certain thing transpired, which was ordering them to the guard house, when they did come in, that by a look from Kid, each man, knew what to do, which they did. In each man jumping for their arms.

[Judge]
Q by J.A.
Did you see anything more of the accused after you were wounded at this time?

[Sieber]
Ans.
No.

[Judge]
Q by J.A.
Did you have direct charge of the accused in your capacity of chief of scouts and did you know him well?

[Sieber]
Ans.
Yes Sir.

[Judge]
Q by J.A.
What influences had the accused had in his band.
The Counsel for the accused objected to this question on the ground that it is leading.

The court was then cleared and closed. After due deliberation the court was reopened, the accused and counsel resuming their seats, and the Judge Advocate announced that the objection had been sustained.[21]

[Judge]

---

[21] This is the first objection that has been sustained that was made by Lt. Baldwin, Defense Counsel. [Sutton]

Q by J.A.
Who was the most prominent man in the band of the accused?

[Baldwin]
Counsel for the accused.
I object. On the grounds that the question is irrelevant, immaterial and incompetent

Page 60 (3716)
[Judge]
The Judge Advocate replied that in the testimony of the previous witness, Captain F.S. Pierce, on almost the identical question had been allowed to be asked by the court. And the Judge Advocate ha not been able to get as full and complete an answer as he felt sure the witness now before the court could give, and that he considered it material to prove that as the accused belonged to the same band as the parties who did the firing, that the latter had induced them to come to the agency and assist him in an escape, if necessary, even to firing on the Commanding Officer. In order to show this it must be proved that the accused was a prominent and influential man in his band.

Then moved and seconded that the court adjourn to meet tomorrow at 8 A.M. The motion to adjourn was lost on vote of the court. Counsel for accused then replied: I intend to show further on that for an nearly nine years the accused has been a soldier of the United States. He has severed his tribal relations. Further I write attention to the fact that you are trying a duly enlisted soldier and not the accused as an indian. The court was then cleared and closed After due

Page 61 (3717)
deliberation the court was reopened the accused and counsel resuming their seats and the Judge Advocate announced that the objection had not been sustained.

[Baldwin?]
The accused then answered:
Kid is fully as prominent as any member in his band.
A motion was then seconded to adjourn to meet tomorrow morning at 8 o'clock. Then moved and seconded that the court take a recess until half past one o'clock.
The court at 11:50 A.M. took a recess to meet again at half past one o'clock. The court met at 1:30p.m. and proceeded to business all the members the Judge Advocate, accused, and counsel present owing to the fact that the witness Albert Siebers was very sick with a wounded leg and the surgeon in attendance having decided that he could not give any more testimony at present, the court proceeded to read the record of the previous day and other business before it. The record of the previous days proceedings

was then read and approved by the court. There moved and seconded that the court adjourn until tomorrow at 8 A.M.[22]

Page 62 (3718)

The court was then at 3:40 P.M. adjourned to meet tomorrow at 8 A.M.

L.D. Tyson
2d Lieut 9th Infantry
Judge Advocate

DAY 5

5th Day.
The Court met pursuant to adjournament at 8:13 o'c a.m.
Present:
1. Major Anson Mills, 10th Cavalry
2. Captain R.G. Smither, 10th Cavalry
3. 1st Lieut. L.P. Hunt, 10th Cavalry
4. 1st Lieut. H.DeLany, 9th Infantry
5. 2nd Lieut. J.B. Hughes, 10th Cavalry
6. 2nd Lieut. W.G. Elliot, 9th Infantry
    And 2nd Lieut. L.D. Tyson, 9th Infantry, Judge Advocate

    Absent
1st Lieut. R.D. Read, Jr., 10th Cavalry
2nd Lieut, C.P. Johnson, 10th Cavalry

The accused present, Counsel absent.
Lieut. Johnson took his seat on the Court at 8:14 o'c A.M.
Lieut. Read and Counsel for accused took their seats at 8:14 ½ o'clock a.m.

The proceedings of the previous day were then read and approved by the Court.
The Judge Advocate announced that he would waive the right to further examine the witness Sieber on the direct examination.

Frederick Knipple a witness for prosecution then

Page 63 (3719)
appeared before the Court and having been duly sworn by the Judge Advocate testified as follows:

---

[22] I think it is interesting that Sieber gave his direct testimony and then he apparently was so ill from his injury his wound that cross-examination had to be postponed for four days. [Rosen]

[Judge]
Q by. J.A.
State your name, occupation and place of residence?

[Knipple]
Issue Clerk, San Carlos, A.T.

[Judge]
Q by J.A.
Do you know the accused and if so as whom?

[Knipple]
Ans.
His name is Kid, He was 1st Sergeant Company "A" Indian Scouts.

[Judge]
Q by J.A.
Where were you on June 1st 1887?

[Knipple]
Ans.
Here at San Carlos, A.T.

[Judge]
Q by J.A.
State if any thing unusual occurred here on that day and if so state all that you saw at the time?

[Knipple]
Ans.
On the evening of June 1st I saw a number of Indians mounted they were going toward Mr. Sieber's tent and I saw Mr. Sieber at Mr. Hortons Store. When I saw Mr. Siever go over to Mr. Antonio Diaz's hosue and from there he came over to Captain Pierce's quarters. Then I saw Antonio Diaz come up towards Sieber's tent. Then I saw Captain Pierce come up to Mr. Sieber's tent. I came up then to Mr. Sieber's tent right afor Captain Pierce. When I got to Sieber's tent I saw some Indians standing in front of this tent, I was just about to go in under the arbor around the tent when I saw an Indian with a belt in his hand. Then there was some indian made a rush to get one of the guns that was standing against the arbor, and I heard Antonio Diaz say, look out for yourselves they are going to shoot. Immediately after I heard Captain Pierce say, look out Sieber they

Page 65 (3720)

56

are going to shoot Siever said let's get there [sic] guns inside and about that time I thought there was going to be trouble and I went into Mr. Siebers tent. Then shooting commenced outside of the tent. MR. Sieber was in the tent and he went outside to fire a shot. He shortly afterwards came back and crawled in the door and said he was shot.

[Judge]
Q by J.A.
Who were those Indians whom you saw coming twards the tent at the time mentioned by you.

[Knipple]
Ans.
I don't know who they were.

[Judge]
Q by J.A.
Did you see any Indians at the tent when you came up whom you knew and if so state who they were?

[Knipple]
Ans.
I think I saw this man Kid but I am not certain, I saw no other Indians whom I knew.

[Judge]
Q by J.A.
What did you go down to Sieber's tent for at this time?

[Knipple]
Ans.
Captain Pierce called me. I sthink he called me to interpret for him.

Cross examination

[Baldwin]
Q by acc.
At the time of the supposed occurrence at the tent, were you alarmed and excited?

[Knipple]
Ans.
Yes, Sir.

[Baldwin]
Q by acc.

When you went into the tent, where was the Captin and where was Sieber?

[Knipple]
Ans.
I think they were just outside of the tent.

[Baldwin]
Q by acc.
When you first went to the tent about how many Indians were at or about that place and were any Scouts other than the accused among the number

Page 65 (3721)
and if so, how many?

[Knipple]
Ans.
I don't know how many. There may have been twenty-five or thirty in all. I don't know—I did not see any that I knew were ____.

[Baldwin]
Q by acc.
Do you understand the apache language?

[Knipple]
Ans.
No Sir.

[Baldwin]
Q by acc.
Are you positive you saw one Indian rush for his gun?

[Knipple]
Ans.
Yes Sir, I am.

[Baldwin]
Q by acc.
Who was that Indian?

[Knipple]
Ans.
I don't know Sir.

[Baldwin]
Q by acc.
How long have you known Kid?

[Knipple]
Ans.
I got to know him since I have been at San Carlos, I have been here about nine months.

[Baldwin]
Q by acc.
Did you see any one short?

[Knipple]
Ans.
No Sir.

Recross Examination

[Judge]
Q by J.A.
Where was Sieber when you saw this Indian rush for his gun?

[Knipple]
Ans.
Mr. Sieber was very close to the gun. About three feet I should think.

Captain A. H. Bowman, a witness for the prosecution then appeared before the court and being duly sworn by the Judge Advocate testified as follows:

[Judge]
Q by J.A.
State your name, rank and station.

[Bowman]
Ans.
   A. H. Bowman, Captain 9th Infantry, San Carlos, A.T.

[Judge]
Q by J.A.
Do you know the accused and if so as whom?

Page 66 (3722)

[Bowman]
Ans.
I recognize him as 1st Sergeant Kid, Indian Scout.

[Judge]
Q by J.A.
Where were you on June 1st 1887?

[Bowman]
Ans.
At this place.

[Judge]
Q by J.A.
State if any thing unusual occurred here on that day and what facts you may know about such occurrences?

[Bowman]
Ans.
There was a disturbance among the Indians, firing upon the Commanding Officer and Al Sieber Chief of Scouts in the vicinity of Siebers tent on thereabouts. The Indians that were a party to this disorder which occurred at or about retreat, I had noticed a half an hour or possibly three quarters of an hour before coming across the parade. They halted in front of Siebers tent. I walked up to where they were, noticing that the Commanding Officer was not there at that time. I walked among them and observed them closely, I noticed that they were armed, all had rifles except on person who was a squaw. After remaining there a very little while I returned to my tent about a hundred yards away, and spoke to Lieut. Watson and others that were near by about these Indians, who were mounted and armed, and then after a little while I returned again to these Indians at AL Siebers tent. I walked around them observed their horses and their arms and on this second occasion I noticed the accused who was one of the party. At this second visit among the Indians I endeavored to get some information from one of them as to who they were, what they were doing there and what was their business. I found the Indians uncommmunicative

Page 67 (3723)
And sullen in their manner towards me and not disposed to talk. On returning to my tent as Captain Pierce and Sieber approached from an opposite direction, I had been in the vicinity of my tent only a few moments when the firing began. I made a movement to go towards the schoolhouse, then returned, when in this position I saw the Indians scampering away from Siebers tent and firing, I then went into my own tent for the purpose of getting a gun; by the time I got out they were pretty well scattered and scampering out on the plain with promiscuous firing in the general direction of the tent.

[Judge]
Q by J.A.
Did you see any individual whom you knew fire a shot at this time?

[Bowman]
Ans.
No.

[Judge]
Q by J.A.
Did you notice any other scouts besides the accused among the number before Siebers tent at the times you mention as having been there?

[Bowman]
Ans.
Yes, I noticed three possibly four others whom I recognized as Scouts on my second visit, I am not positive that more than two of them were Scouts as I could not identify them. Five of the whole party had gov't guns, defautry rifles, three or four others Winchester rifles, all there arms in excellent order. One a Cavalry Carbine.

[Judge]
Q by J.A.
Did you see any one about Siebers tent wounded?

[Bowman]
Ans.
I saw Al Sieber himself wounded and a good deal of blood flowing from his leg when I walked to his tent a few minutes after the Indians had left.

Cross Examination

Page 68 (3724)
[Baldwin]
Q by acc:
Have you not expressed bias and prejudice against this defendant when a member of this Court?

[Bowman]
Ans.
Yes.

[Baldwin]

61

Q by acc.
You say Indians fired upon Captain Pierce, the Commanding Officer, did you see anyone do so?

[Bowman]
Ans.
Any individual Indian, No. I saw firing in the direction of Siebers tent, this firing was done by the Indians in front of the tent, before alluded to.

[Baldwin]
Q by acc.
Did you see Captain Pierce at the tent until after the alleged occurrence?

[Bowman]
Ans.
I saw him there before the firning began, and I also saw him there afterwards.

[Baldwin]
Q by acc.
Where were you, when you say you saw him at the tent?

[Bowman]
Ans.
I was in front of my own tent a hundred yards away, after the firing I saw him at Siebers tent.

[Baldwin]
Q by acc.
How far in front of your tent on the first occasion?

[Bowman]
Ans.
I was somewhere in the vicinity of my tent. I saw Captain Pierce several times, I walked down among the Indians and returned towards my tent, stopped several times. Each time I saw Al Sieber and others I saw him when I was somewhere in the vicinity of my tent apparently engaged with the Indians.

[Baldwin]
Q by acc.
You have said you saw Captain Pierce at Siebers tent, when you were in front of your tent, now please state, how far in front of your tent you

Page 69 (3725)

62

were when you then saw him, and if any other persons were present near you?

[Bowman]
Ans.
For the word "front" I ought to use the word vicinity. There were several gentlemen in the neighborhood.

[Baldwin]
Q by. Acc.
The first time you went to Siebers tent, about how many Indians were there present?

[Bowman]
Ans.
I counted elevent. There were the Indians and one woman.

[Baldwin]
Q by Acc.
The second time, about how many Indians were there present?

[Bowman]
Ans.
There were about the same number.

[Baldwin]
Q by acc.
Were your visits to this tent accuated by curiosity or in anticipation of trouble?

[Bowman]
Ans.
The attitude of these Indians impressed me as being extraordinary and unusual. I saw nothing of Captain Pierce or Al Sieber when first looking up there and went there for the purpose of satisfying myself about them.

[Baldwin]
Q by acc.
Please give the result of that investigation in respect to the impressions as to trouble made upon your mind?

Page 70 (3726)

[Bowman]
Ans.

I was impressed with the demeanor and bearing of these Indians, but having no direct authority or responsibility my deductions and conclusions were merely speculations in my own mind as to what was to happen.

[Baldwin]
Q by Acc:
Did you believe at that time there was trouble brewing at that tent?

[Bowman]
Ans.
No. I only thought that something unusual was in the air.

[Baldwin]
Q by Acc:
Did that something unusual have any relation to subsequent alleged events and did you communicate any impressions made by the sullen manner you mention by these Indians, to the Commanding Officer.

[Bowman]
Ans.
They did have connection undoubtedly, with what happened afterwards. I made no communication to the Commanding Officer for the double reason that I was not sufficiently advised of Indian Affairs to presume to make any asuggestions or offer amy opinions about matters that related purely to indian Affairs, and for another reason I did not have an opportunity in consequence of the disorders following a few moments after my observation.

[Baldwin]
Q by Acc:
Can you speak and understand the Apache language?

[Bowman]
Ans.
No. I allude to speaking to them in my testimony. The man I spoke to had

Page 71 (3727)
a white hat on and was a very light complexioned man; He having his hair out short had the appearance of a boy who had been at school, or a Mexican, and I spoke to him with a view of finding out what they wanted there.

[Baldwin]
Q by Acc:
Were you armed the second time you went to Siebers tent?

64

[Bowman]
Ans.
No.

[Baldwin]
Q by Acc:
Was the person you spoke to dressed in the garb of an Indian?

[Bowman]
Ans.
Yes. With the exception of the hat.

[Baldwin]
Q by Acc:
Do you know when you spoke to this Indian or Indians, at the tent whether he or they could understand you?

[Bowman]
Ans.
No.

[Baldwin]
Q by Acc:
Then please explain as you can-not understand Apache, or know of a party of Indians who can not understand you, when you say these Indians were uncommunicative and not disposed to talk?

[Bowman]
Ans.
They shook their heads and said: "No savey."

[Baldwin]
Q by Acc:
You say you observed these Indians closely, did they have belts and ammunication on their persons.

[Bowman]
Ans.
They did. Plenty of it.

[Baldwin]
Q by Acc:

What kind of belts and ammunitions

Page 72 (3728)
[Bowman]
Ans:
Five of them had the old fashioned prairie belt or something similar, balance had nondescript belts made of canvass or something.

[Baldwin]
Q by Acc:
Did you see Captain Pierc at Sievers tent during the alleged firing?

[Bowman]
Ans.
I saw him there a moment before and afterwards.

[Baldwin]
Q by Acc:
Do you know of your own knowledge against whom this firing mentioned in evidence, was directed?

[Bowman]
Ans.
No. Not further than that it was directed at the parties in front of that tent.

[Baldwin]
Q by Acc:
During the melee, did you see any persons firing , and if so, were not some of those persons members of C.A. Indian scouts, and other than the five scouts you have mentioned?

[Bowman]
Ans.
I did see one Indian firing at them after they had gotten away some distance.

William Dudin a witness for the prosecution then appeared before the court, was then duly sworn by the Judge Advocate, and testified as follows:

[Judge]
Q by J.A.
State your name, occupation, and place of residence.

[Dudin]

Ans.
William Dudin, blacksmith San Carlos, A.T.

Page 73 (3729)
[Judge]
Q by Acc:
Do you know the accused, and if so as whom?

[Dudin]
Ans.
I know him as Kid. He was first sergeant of the Indian Company

[Judge]
Q by J.A.
Where were you on June 1st 1887.

[Dudin]
Ans.
I was here at San Carlos.

[Judge]
Q by J.A.
State if anything unusual occurred here on that day, and if so state all that you saw of these occurrences. Before this question was answered it was moved and seconded that the court take a recess until 1:30 P.M. The court then at 12:00 P.M took a recess until 1:30 P.M. The court then at 1:30 P.M. resumed its proceedings all the members of the court the Judge Advocate, accused, and counsel, being present.
The witness then answered:

[Dudin]
Ans.
I don't think anything occurred on the first of June. To the best of my memory it was the second

[Judge]
Q by Acc:
State if anything unusual occurred on or about the first of June, and is so state all that you can of these occurrences?

[Dudin]
Ans.

I did not see anything until in the evening. I saw this "Indian" the prisoner and seven or eight m ore come in and go across to Siebers tent, About fifteen minutes after that I heard the shooting. This was about

Page 74 (3730)
all until I saw this prisoner come between this outhouse and the dormitory, (pointing in the direction of the houses mentioned) and fire a shot. This man here then went down the hill and got on his mule, an iraw [?] gray mule and another man got up behind him after he had ridden about ten or fifteen yards and they went off.

[Judge]
Q by J.A.
In what direction did the prisoner shoot?

[Dudin]
Ans.
He shot towards the sutlers store

[Judge]
Q by J.A.
Where is Siebers tent with reference to the place at which you saw the prisoner at the time you saw him shoot, and the Sutler store?

[Dudin]
Ans.
It was to the left of the way he fired the shot.

[Judge]
Q by J.A.
Was it in front or rear of the accused at this time.

[Dudin]
Ans.
At the time he fired the shot it was in front of him.

[Judge]
Q by J.A.
How far were you from him?

[Dudin]
Ans.
I should judge it was 80 or 100 yards.

[Judge]
Q by J.A.
How far is Siebers tent to the left of a perpendicular line drawn from the tent to a line drawn through the spot where the accused was standing and the Sutler's Store?

[Dudin]
Ans.
I should think it would be thirty-yards.

[Judge]
Q by J.A.
Could you see Siebers tent from where

Page 75 (3731)
you were standing at this time?

[Dudin]
Ans.
No.

[Judge]
Q by J.A.
Could you see the Sutlers Store at that time?

[Dudin]
Ans.
No.

[Judge]
Q by J.A.
Can you say definitey then as you have said that Seibers tent and the Sutlers Store are almost in the same time that the accused was not firing at Siebers tent?

[Dudin]
Ans.
When he pointed his gun I could tell pretty near which way he was shooting.

[Judge]
Q by J.A.
What did the accused fire with?

[Dudin]
Ans.

An infantry gun – "a long Tom".

[Judge]
Q by J.A.
Did you see any one else fire a shot on this occasion?

[Dudin]
Ans.
Yes.

[Judge]
Q by J.A.
State who it was?

[Dudin]
It was a Scout, his name is John. He belongs here and is a Corporal I think. I think he is now acting 1st Sargeant, He was over at the hospital and fired towards the Gila at the renegades as I supposed.

[Judge]
Q by J.A.
State the names of all the persons whom you know to have fired shots on this occasion?

[Dudin]
Ans.
This man here (pointing to the prisoner) and John

[Judge]
Q by J.A.
Is the accused well known to you by sight?

[Dudin]
Ans.
Yes, Sir.

[Judge]
Q by J.A.
At about what hour did all these occurrences take place?

Page 76 (3732)
[Dudin]

70

Ans.
I should judge it was about 6 P.M.

[Judge]
Q by J.A.
How far was the accused from the Sutlers Store when he did the firing mentioned?

[Dudin]
Ans.
I think it must be a thousand yards.

CROSS EXAMINATION
[Baldwin]
Q by Acc:
At the time you mention, how was the accused dressed?

[Dudin]
Ans.
About like he is at present.

[Baldwin]
Q by Acc:
Did he have a blouse on with Chevrons?

[Dudin]
Ans.
No.

[Baldwin]
Q by Acc:
Please point out the exact spot where the accused stood when he fired the shot you mention?

[Dudin]
Ans.
I don't know that I can point it out exactly, but it was somewhere between that outhouse and dormitory (pointing to the outhouse and dormitory in question.)

[Baldwin]
Q by Acc:
How far on this side of that outhouse was he?

[Dudin]

71

Ans.
I don't think he was any this side of it, I think he was just about between the outhouse and dormitory.

[Baldwin]
Q by Acc:
Were you at the Blacksmith shop at this time?

[Dudin]
Ans.
No, I was this side of it, I was close to it.

[Baldwin]
Q by Acc:
Did you see the face of the person who fired the shot?

[Dudin]
Ans.
Yes.

[Baldwin]
Q by Acc:
Did you see the face of the person when the shot was fired?

Page 77 (3733)
[Dudin]
Ans.
No, I did not.

[Baldwin]
You have said the accused got a mule, where did he go to get that mule?

[Dudin]
He went down to the Scouts Corral.

[Baldwin]
Q by J.A.
How far off from you was that corral and did you see him get the mule?

[Dudin]
Ans.
I must have been between a hundred and a hundred and afifty yerds from where I stood. He got the mule.

[Baldwin]
Q by J.A.
When this shot mentioned, was fired, were any other Indians present near the person who fired that shot?

[Dudin]
Ans.
Not that I know of, I didn't see any.

[Baldwin]
Q by Acc:
Before and after this shot was there much firing, and did any other Indians pass you,from the same direction you have mentioned?

[Dudin]
Ans.
There was firing before this and there were seven or eight Indians who passed afterwards.

[Baldwin]
Q by Acc:
Please give the names of persons, who were near you , at the time, if any?

[Dudin]
Ans.
I don't know the names. There was one Corporal George Harvey I believe, there were some Infantry men there that I don't know the names.

[Baldwin]
Q by Acc:
State the position the person took when he fired the shot mentioned in the direction of Traders Store, and when at the outhouse;

[Dudin]
Ans.
He was facing towards the Traders Store with his back towards me, He was standing when he fired.

Page 78 (3734)
[Baldwin]
Q by Acc:
In point of time, how long after this shot did John fire?

73

[Dudin]
Ans.
I must have been ten or fifteen minutes.

[Baldwin]
Q by Acc:
Is your mind unbiased and unprejudiced against this defendant and have you any interest in the result of this trial?

[Dudin]
Ans.
The court was then cleared and closed. After due deliberation the court was reopened, the accused and Counsel resuming their seats.

[Baldwin]
Q by Acc:
I do not like to answer that question. It is unbiased as far as this trial is concerned. I can't say that I have any interest in the result of this trial. I can't say that I have any prejudice against this accused.

[Baldwin]
Q by Acc:
Upon your oath, do you positively swear that you saw, the accused fire a shot on the day in question and from any part of his apparel could you recognize him?

This question was objected to by a member.

The Counsel had nothing to say.

The Court was then cleared and closed. After due deliberation the court was reopened, the accused and council resuming their seats and the President announced that the objection was withdrawn.

[Dudin]
Ans.
Yes, Sir.

[Baldwin]
Q by Acc:
Mention the part of his apparel that you recognized?

[Dudin]

Ans.
He was dressed in a similar way to what

Page 79 (3735)
he is now.

[Baldwin]
Q by Acc:
Is not the present dress of the accused similar to that of other Indians?

[Dudin]
Ans.
Yes, It is similar, some have red shirts and some white ones, drawers and things

[Baldwin]
Q by Acc:
They so far as dress and face are concerned you did not recognize the accused at the time, this particular shot was fired, being at the time about one hundred years away?

[Dudin]
Ans.
Yes. I recognized his gait, walk and face, I have seen him so foten, I know him better than I know you (referring to the Judge Advocate).

RE-EXAMINATION

[Judge]
Q by J.A.
Did you take particular notice of the accused after he turned around after firing at the time mentioned.

The counsel for accused then made the following objection to the question.

Upon the ground that the question does not, clear up any dubious meaning of the witness, but on the contrary opens up the direct examination or examination in chief, and thereby prevents as it now stands, the counsel from cross examining upon any new matter brought out in answer.

The Judge Advocate replied that he believed this question would clear up all doubts as to the identity of the man who fired the shot in question.

The court was then cleared and closed. After due deliberation the court was reopened, the accused and counsel resuming their seats and the Judge Advocate answered that the objection has not been sustained.

[Dudin]
Ans.
Yes, Sir.

[Baldwin]
Please state fully how you came to devote so much particular and special attention to the accused on this day?

Page 80 (3736)
[Dudin]
Ans.
I heard the shooting and he was the first man I saw going over the hill, towards the corral.

[Court]
Q by Court:
Did you see where the ball from Kid's gun struck the ground?

[Dudin]
Ans.
No sir.

[Court]
Q by Court:
Was any one in this line of fire if so who?

[Dudin]
Ans.
I didn't see any one.

[Court]
Q by Court:
Are you positive that the occurrences you speak of took place on the second of June?

[Dudin]
Ans.
No sir. I am not. I thought it was the second. I am not positive.

[Court]

Q by Court:
Will you explain how it was that you could see a shot fired in the direction of the traders store, and why you could not see the points indicated as being in the vicinity of the line of ifre or why you could not see any one that may have been in the supposed line of fire?

[Dudin]
Ans.
The accused was right on the brow of the hill and I was below – I could see from the way he was pointing his gun the direction he was firing, because I had been over the ground so often.

Private Chas S. Chew-Co. A. 9th Infantry, a witness for the prosecution then appeared before the court and having been duly

Page 81 (3737)
Sworn by the Judge Advocate testified as follows:

[Judge]
Q by J.A.:
State yourname, rank and station?

[Chew]
Ans.
Chas. S. Chew, Private Co. A 9th Infantry San Carlos, Arizona.

[Judge]
Q by J.A.
Do you know the accused and is so as whom?

[Chew]
Ans.
No sir. I don't know him.

[Judge]
Q by J.A.
Where were you on June 1st 1887?

[Chew]
Ans.
I was working in the Quartermaster's Office all day at San Carlos.

[Judge]

Q by J.A.
Did any thing unusual occur here on that day?

[Chew]
Ans.
Yes sir.

[Judge]
Q by J.A.
Did you hear or see any firing here on that day, if so state where?

[Chew]
Ans.
Yes sir. I heard shots fired and saw one.

[Judge]
Q by J.A.
State where, when and by whom, this shot was fired?

[Chew]
Ans.
I could not tell. It was an Indian that fired it. It was right on the brow of this hill. This hill by this water closet here. (pointing to a water closet close by)

[Judge]
Q by J.A.
How far was the Indian from the outhouse near the water closet?

[Chew]
Ans.
I should judge from where I was he was about ten yards.

Page 82 (3738)
[Judge]
Q by J.A.
About what time was this?

[Chew]
Ans.
It was about a quarter after 7 P.M. I should judge.

[Judge]
Q by J.A.

78

You say you could not tell who fired the shot; do you know how the person was dressed?

[Chew]
No sir. I did not take particular notice.

[Judge]
Q by J.A.
What did the person in question fire with?

[Chew]
Ans.
I thought it was a carbine.

[Judge]
Q by J.A.
How far were you from the person who fired the shot?

[Chew]
Ans.
Between sixty and seventy yards.

[Judge]
Q by J.A.
In what direction did he fire it?

[Chew]
Ans.
He fired up towards Siebers tent.

[Judge]
Q by J.A.
You say you heard shots fired; how long after you heard these shots fired first, was it before you saw the person in question fire the shot?

[Chew]
Ans.
I should judge about a half a minute.

[Judge]
Q by J.A.
In what direction did the person who fired the shot that you saw go after he had fired?

[Chew]
Ans.
He started right down over the hill and passed between the two corrals.

Page 83 (3738)
[Judge]
Q by J.A.
Did you see him again?

[Chew]
Ans.
No sir.

[Judge]
Q by J.A.
Could you see the scouts corral from where you were standing?

[Chew]
Ans.
Yes sir.

[Judge]
Q by J.A.
Did you recognize the person who fired the shot, to be a scout or not?

[Chew]
Ans.
No sir. I could not recognize scouts or other Indians at all.

[Judge]
Q by J.A.
Was the person in question between you and the water closet, or was he between the watercloset and the dormitory?

[Chew]
Ans.
He was between myself and the water-closet.

[Judge]
Q by J.A.
Where were you standing at this time?

[Chew]

Ans.
At the end of the commissary tent.

[Judge]
Q by J.A.
Did the person in question wear any military apparel?

[Chew]
Ans.
No sir not that I noticed.

Then moved and seconded that the court adjourn until tomorrow at 8:30 A.M. on vote of the court the motion was lost.

CROSS EXAMINATION

[Baldwin]
Q by Acc:
State where the firing, other than the particular shot mentioned took place?

[Chew]
It was up this way some where

Page 84 (3740)
I could not tell exactly. I did not get out quick enough.

[Baldwin]
Q by Acc.
How long had the firing mentioned been going on before you come out?

[Chew]
Ans.
About half a minute.

[Baldwin]
Q by Acc.
Had there been any other firing before this?

[Chew]
Ans.
No sire, not before that time.

[Baldwin]

Q by Acc.
After you came out, did you observe any Indians pass in front of you, or within a limit of sixty or seventy yards?

[Chew]
Ans.
Yes, sir.

[Baldwin]
Q by Acc.
Where did they come from?

[Chew]
Ans.
They came down from in front of the officers tent.

[Baldwin]
Q by Acc.
About how near to the tents?

[Chew]
Ans.
I should say thirty or forty years.

[Baldwin]
Q by Acc.
Did this shooting and the movement of the Indians receive your attention?

[Chew]
Ans.
Yes sir.

[Baldwin]
Q by Acc.
Where did the Indians who passed in front of the officers tent go?

[Chew]
Ans.
They all went over the hill.

[Baldwin]
Q by Acc.
When this person fired the shot was he kneeling or standing?

[Chew]
Ans.
He was mounted.

[Baldwin]
Q by Acc.
As there was firing, and Indians

Page 85 (3741)
Passing by you, how came you to give the particular person who fired the shot so much attentions?

[Chew]
Ans.
Because he was the only person I saw fire.

[Baldwin]
Q by Acc.
Where did you go after the occurrence?

[Chew]
Ans.
I did not go anywhere I stayed right there.

[Baldwin]
Q by Acc.
When you came out of the tent did you see any Indians, other than the one who fired the shot mentioned by you?

[Chew]
Ans.
Yes sir.

Then moved and seconded that the court sdjourn until 8:30 am. The court then at 5:30 pm adjourned until tomorrow 8:30 a.m.

J.D. Tyson
2nd Lieut 9th Infantry
Judge Advocate

6th DAY

San Carlos A.T.
July 5, 1887

The court met pursuant to adjournment at 8:30 am

Present:

1. Major Anson Mills, 10th Cavalry
2. Captain R.G. Smither, 10th Cavalry
3. 1st Lieut. L.P. Hunt, 10th Cavalry

Page 86 (3742)

4. 1st Lieut. H. DeLaney, 9th Infantry
5. 1st Lieut. R.D. Read, 10th Calvary
6. 2nd Lieut. C.O. Johnson, 10th Calvary
7. 2nd Lieut. J.B. Hughes, 10th Calvary
8. 2nd Lieut. W. G. Elliot, 9th Infantry and
   2nd Lieut. L.D. Tyson, 9th Infantry, Judge Advocate

Accused and counsel present.

The proceedings of the previous day were then read and approved.

The court then permitted the Counsel for accused to have the following appear upon the record.

[Baldwin]
I request that it appear upon the record, that owing to the physical condition of Mr. Sieber, and who was waiting to have an operation performed upon him by attending surgeons, that the cross examination by counsel, could not be had at the time, and conclusion of the examination in chief.

"Tony" 1st Sargt Company "A" Indian Scouts then appeared before the Court and having been duly sworn by the Judge Advocate, testified as follows: Counsel objected on following grounds:

[Baldwin]
I object to the swearing of this witness, on the ground of incompetency by reason of want of religious principle. The Court was then cleared and closed, after due

84

deliberation the court was reopened, the accused and counsel resuming their seats and the Judge Advocate announced that the objection is not sustained:

[Judge]
Q by J.A.
State your name, rank and station?

[Tony]
Ans.
"Tony" 1st Sergeant Indian Scourts, San Carlos A.T.

[Judge]
Q by J.A.
Do you know the prisoner and if so state

Page 84, (3743)
his name?

[Tony]
Ans.
His Indian name is Ha-hau-au-tell, White men call him "Kid"

[Judge]
Q by JA
Where were you on the 1st of June 1887?

[Tony]
Ans.
At San Carlos. I was at my tent.

[Judge]
Q by JA
Did you see or hear any firing that day, and at what time?

[Tony]
I heard a shot about sundown.

[Judge]
Q by JA
About how many shots did you hear and what did you do?

[Tony]
Ans.

I don't know, I hard shots several times, I went to get my gun at this time.

[Judge]
Q by JA
Do you know who did this firing and if so state who it was?

[Tony]
Ans.
I was running from them. I don't know who did it.

[Judge]
Q by JA
Did you do any firing and if so state at shat or shom you were firing?

[Tony]
Ans.
I did fire at the Indians but I knew only ont that was with them. I fired at the renegades.

[Judge]
Q by JA
How many shots did you fire?

[Tony]
Ans.
I fired twice. The first fire I made there were three men standing together. I knew only one of them.

[Judge]
Q by JA
State the name of this one you knew?

[Tony]
Ans.
His name is Bachaaudah.

[Judge]
Q by JA
Did you see any other persons fire on the renegades at the time mentioned and if so stat their names?

[Tony]
Ans.
One of the scouts fired at the renegades, His name

Page 88 (3744)
is John.

[Judge]
Q by Court
Did you see any body fire besides yourself and John? Who were they and at whom did they fire?

Counsel objected on following grounds:

I object. For the reason, that the government has confided its interests to the Judge Advocate, and for members of the Court to invade the duties of the Judge Advocate, by originating evidence, effects the rights of the accused. [23]

The Court was then cleared and closed, after due deliberations the Court was reopened, the accused and counsel resuming their seats and the Judge Advocate announced that the objection has not been sustained.

[Tony]
Ans.
No, sir.

W.B. Horton of San Carlos A.T. was then duly sworn as interpreter.

Antonio Diaz, a witness for the prosecution was then duly sworn by the Jduge Advocate and testified as follows

[Judge]
Q by JA
State your name, occupation and place of residence?

[Antonio]
Ans.
Antonio Diaz, Interpreter, San Carlos, A.T.

[Judge]
Q by JA
Do you know the accused, and if so as whom?

---

[23] For this reason stated by the Defense Counsel, Lt. Baldwin, questions from the Members must now be in writing and submitted to the Judge who will then ask the question, if proper.

[Antonio]
Yes, "Kid"

[Judge]
Q by JA
Where were you on June 1st 1887?
[Antonio]
Ans.
I was here at San Carlos.

[Judge]
Q by JA
Sstate if anything unusual occurred here on that day?

[Antonio]
Yes.

[Judge]
Q by JA
State where you were at the time of these

Page 89 (3745)
Recurrences, and all that you saw and heard of them?

[Antonio]
Ans.
I was with Captain Pierce Fred and Al Sieber at Al Siebers tent. When the Indians came in the direction of the tent, Mr. Sieber went to my house and called me. After he had called me in about five minutes I came from my house, thinking that Mr. Sieber was at this tent, then I came to Al Siebers tent and the Indians were there, When I arrived where they were, I said to them how are you. Then I directed my attention to Kid. At the time I directed my attention to him he held out his hand to me. Then I said to him, how is it you have been out so long. Kid replied that we have been off and have killed a man at the Aravaipa, Then I said to him I am sorry for it. Then he replied to me, it is no matter of yours nor of the agents. It is our affair and no one elses. Then I said to him that is all right, it does not matter to me whether you kill ten or a dozen Indiains a day, and I suppose it does not matter to the Captain either. Then I said to him you ought not to have done that. You ought to have considered that you have a family. I said to him those that have children and families should consider a misfortune that may overtake them, three of those men that went out with you were of mature years and should have repented. I said to thim there is Migree, Askisalaha and Catchiklu, that they are of mature age and should have advised

88

Page 90 (3746)
You who are younger not to have committed this act, but to have returned. I said to him how is it that you have gone forty miles from here killed a man in cold blood and returned here? Is is not our custom to kill people, thus. Yes said "Kid" but we did it. It is nobody's business. Then nothing else occurred. He (pointing to the Accused) alone was speaker, and during this time none of the others spoke. Then Mr. Siebers arrived and when he arrived there he said "hello Kid" and Kid said "hello". Al Sieber opened the door of his tent and entered. He as probably in his tent three minutes when Sieber came out. In as short time after Captain Pierce arrived and Fred also arrived, then I stepped to one side under the arbor. Then the Captina said to Sieber, "Let them give up their arms." Then Sieber asked them for their arms, and the first one that gave up his arms was Kid. Then I noticed that two others gave up their arms, but during this time my attention was fixed on the Indians outside, and while they were giving up their arms, Captain Pierce turned to John the scout and said, get ready to take these men to the calaboose, when the Captain said this, the Indian Bashandoth stepped forward to get his belt. Then I suspected mischief. The Captain said to him what are you going to do with that belt? The Indian

Page 91 (3747)
Replied "nothing," but he made a motion to take off the knife scabbard and instead of doing so he put the belt around him. I was particularly watchful of the Indians who were seated in front of the tent with their arms in their hands. At the moment that the Indians fastened the belt around him he grabbed his gun and Bachoandoth jumped like a flash of lightning and to one side, and he said, "there is nothing for us to do but to fight," and during this time "Kid" was still standing there. When Bachaoandoth made this remark, I looked off to one side and said to Fred, "look out they are a going to shoot you in the back." I was facing the Indians in front of the tent. Mr. Sieber the Captain and Fred had their backs turned to them. I didn't say anthing more. Then I fell down going backwards. When I threw myself back the Indians fired shots through the side of the tent. I heard the voice of an Indian say and I don't know what Indian it was, "Let the tropps follow up quickly, we are going." Then I ran in the direction of the traders store, and a shot was fired at me, and I met one of the friendly scouts, and he fired a shot at the retreating Indians. The Indians were just disappearing over the bluff. The Indian who

Page 92 (3748)
Fired the shot was coming along the main road. I don't know who this Indian was. He was a soldier I know. When I was turning around Siebers tent I noticed "Rowdy" fire a shot from near the scouts tent and came running in the direction of the school house. I went over running toward shte scouts camp and I called to themand there were not scouts there. They had all gone away. I then went to my house still running and I ran like a deer.

[Judge]

Q by JA

What were the names of the Five scouts you saw when you came to Seibers tent?

[Antonio]
Ans.
One is named Kid, another Askisalaha, another Mangy another Backoandoth, another Nacauginsay.

[Judge]
Q by JA
Did you see any one else reach for his arms at the time you say Backoandoth took his?

[Antonio]
I did not. My attention was not directed to the Indians who were giving up their arms, my attention was particularly directed to those Indians who were in front.

[Judge]
Q by JA
Who were those Indians in front?

[Antonio]
Ans.
I remember particularly Migell, a Yaqui and Gonshayee. He had no arms. I remember another Indian whose name is Gonsiye and a brother of his.

[Judge]
Q by JA
To what band did these Indians belong?

[Antonio]
Ans.
To Gonshayee's band

[Judge]
Q by JA
To what band does "Kid" belong?

[Antonio]
Ans.
To Gonshayee's band

[Judge]
Q by JA

Did you see any one of these scouts

Page 93 (3749)
Mentioned fire a shot while you were at Siebers tent?

[Antonio]
Ans.
I did not notice any other of these scouts fire in the direction of the tent except Bachondoth, and I don't remember to have seen any others, I noticed him sieze his belt, grab his gun and jump to one side and say all we have got to do si to fight and put a cartridge in his gun.

[Judge]
Q by JA
In what direction did Bachoaudoth fire this shot?

[Antonio]
Ans.
 In the direction of the door of the tent.

[Judge]
Q by JA
Where was Captain Pierce at that time?

[Antonio]
Ans.
He was standing in the door of the tent.

[Judge]
Q by JA
Did you see any one of those Indians outside of the arbor do any firing?

[Antonio]
Ans.
At the time this was Bachoaudoth jumped out there then I saw Migree the Yaqui throw down his arms, (showing the position of firing). He had a Winchester in his hand and fired.

[Judge]
Q by JA
In what direction did Miquel fire?

[Antonio]

Ans.
In the direction of the tent.

[Judge]
Q by JA
Where was Captain Pierce at this time?

[Antonio]
Ans.
He was standing a little in front of the door of the tent.

[Judge]
Q by JA
You say that Bacho audoth took his belt and that you suspected mischief. What was it you saw that suggested mischief?

[Antonio]
Ans.
I suspected mischief because he seized his belt and placed it around him.

Page 94 (3750)

The Court then at 12 oclock n. Took a recess until 1:30 P.M.

The Court at 1:30 P.M. resumed proceedings, all the members, the Judge Advocate, accused and counsel being present.

Cross Examination

[Baldwin]
Q by Counsel
How long have you been interpreter at this agency?

[Antonio]
Ans.
About four years and eight months.

[Baldwin]
Q by Counsel
Are you usually employed when the agent is having conference with the Indians?

[Antonio]
Ans.

Since Captain Pierce took charge when he wants an interpreter he calls on me. I am military interpreter.

[Baldwin]
Q by Counsel
Give the name of the man Killed in the Aravaipa, and relate the causes so far as you know, that caused that act?

[Antonio]
Ans.
He was called "Rip". The cause was that they had a tizwin drunk alone, and then former chief was killed. At this drunk "Rip" and son-in-law of the former Chief had a fight. Then the former Chief went in to separate them; in doing so a nephew-in-law drew his gun and fired and killed the Chief.

[Baldwin]
Q by Counsel
When you were talking to the accused, at Siebers tent, did you, say to him or to any other Indians what would be done with them for Killing this person?

[Antonio]
Ans.
I did not tell him anything on that point.

Page 95 (3751)
[Baldwin]
Q by Counsel
Was "Rip" an Indian?

[Antonio]
Ans.
He was.

[Baldwin]
Q by Counsel
When Bacho-au-doth, said there is nothing for us to do but fight, as you state, did the accused say anthing?

 [Antonio]
Ans.
He said nothing.

[Baldwin]

Q by Counsel
Did he do anything?

[Antonio]
Ans.
I did not pay attention. I took precautions to escape.

[Baldwin]
Q by Counsel
When you said to the accused, How is it you have gon forty-miles, killed a man and returned here tell all, everything he (the accused) said at that time?

[Antonio]
Ans.
He said when I told him, it is no matter to you nor to any of the Americans. It is our affair. "Kid" repeated two or three times it was none of Antonios business. They had gone to kill this man and they had done it.

[Baldwin]
Q by Counsel
When Captain Pierce was standing in the door of Siebers tent how was he Captain Pierce facing?

[Antonio]
Ans.
He was looking and partially facing the tent, having his side to the tent.

[Baldwin]
Q by Counsel
About how many scouts other than the three you mention, were at or about the tent?

[Antonio]
Ans.
There were many Indians. I did not notice that.

Page 96 (3752)
[Baldwin]
Q by Counsel
Is it not usual, when Indians have done anything wrong, to meet at Siebers tent or other places and have a conference with their agent Captain Pierce?

[Antonio]
Ans.

When they have these talks they go where they are desired to have the talks.

[Baldwin]
Q by Counsel
As interpreter did you not understand that this meeting at Siebers tent was for the purpose of having a talk about killing this Indian "Rip" and for having been away on a tizwin drunk?

[Antonio]
Ans.
The Captain did not say anything to me about it and perhaps it was for that object.

[Baldwin]
Q by Counsel
During the occurrences, mentioned by you at the tent, were you not very much excited and very much frightened?

[Antonio]
Ans.
I was excited when the shooting began but before that time I was not excited.
Why should I be frightened?

[Baldwin]
Q by Counsel
As you fell back and did not devote much attention to those under the arbor and near Captain Pierce and Sieber, and then ran around the tent and ran home like a deer, how can you swear on your oath that you saw Bachoandoth shoot at Captain Pierce, and Miguel fire a shot toward the tent?

[Antonio]
Ans.
I suspected something when the Indian took hold of the cartridge belt and the Captain said what are you going to do with it,

Page 97 (3753)

And the Indian replied nothing. Bachoandoth seized the cartridge belt, made a motion as if to take off the knoife scabbard, then he fixed the belt around him immediately, then grabbed a rifle, then he made a jumpt to one side, this I would sear to a thousand times, then he put a cartirdige in the rifle and at the same time he says, "there is nothing fo us to do but to fight," and then brought his gun down in position to fire. I was then observing all the Indians in front, and about the same time two shots were fired, one by Miguel, , the other by Bachoandoth both nearly at the same time.

[Baldwin]
Q by counsel
Did you see them fire?

[Antonio]
Ans.
Yes sir.

[Baldwin]
Q by Counsel
 How did Bachoandoth hold his gun?

[Antonio]
Ans.
In front of his body.

The counsel for defense said that he did not desire to go on with this case until he could have an opportunity to examine Mr. Sieber. The court then proceeded to the trial of other cases before it until such time, as Mr. Sieber could be further reexamined.[24]

L.D. Tyson
2nd Lieut, 9th Infantry
Judge Advocate

Page 98 (3754)
Official copy of telegraphic order rec'd this day Juy 5th, 1887.

[READS as follows]
Headquarters Department of Arizona
                        Los Angeles, Cal., June 30, 1887
Special Orders }
No. 67   }

(Extract.)

---

[24] I am surprised that they gave the Defense Council the opportunity to cross-examine Sieber but apparently they were cognizant of the fact that the defense, the Kid, had the right to cross-examine the witness and they in fact gave the defense counsel the right to cross-examine Sieber. I mean, I have never seen anything like it but I assume that is what would happen, otherwise you would have to strike his direct testimony. I do not know how else you can do it. You have a right to confrontation. [Rosen]

1. Major Anson Mills, 1st Lieutenants L.P. Hunt and R.D. Read, Jr., 10th Calvary, are detailed as members of the general court martial convened at San Carlos, A.T., by Field Orders, No. 1, c.s.

Upon completion of the cases before the court they will return to their stations.

The journeys as directed are necessary for the public service.

By command of Brigadier General Miles:

M. Barber,
Assistant Adjutant General

7th DAY   [four days later]

San Carlos, A.T.
July 9, 1887

The court met pursuant to adjournment at 8:30 A.M.

Present:

1. Major Anson Mills, 10th calvary
2. Captain R.G. Smither, 10th Calvary
3. 1st Lieut. L.P. Hunt, 10th Calvary
4. 1st Lieut. N DeLany, 9th Infantry
5. 1st Lieut. R.D. Read, 10th calvary
6. 2nd Lieut. C.P. Johnson, 10th calvary
7. 2nd Lieut. J.B. Hughes, 10th Calvary
8. 2nd Lieut. W.G. Elliott 9th infantry, and
2nd Lieut. L.D. Tyson, 9th Infantry, Judge Advocate,
The accused and Counsel present

The court then proceeded to Sibers room to take his testimony on cross examination.

The court having arrived there the witness testified as follows:

[Baldwin]
Q by Counsel
During the eight years you have known the accused, how long has he been a soldier?

[Sieber]
Ans.

He has been a soldier most of the time, In all he must have been a soldier four or five years out of the eight.

[Baldwin]
Q by Counsel
You say Captain Pierce took the arms from the Scouts. Did he use any force in doing so?

Page 99 (3756)
[Sieber]
Ans.
None. They handed their arms to me.

[Baldwin]
Q by Counsel
How many times did the Captain order them to the Calaboose and in what tone of voice did he give the order?

[Sieber]
Ans.
Twice. In an ordinary tone of voice.

[Baldwin]
Q by Counsel
You say the accused gave a look which signified to you for them to rush for their arms, was that your impression at the time or was it not the result of deductions and conclusions formed after the occurrences mentioned?

[Sieber]
Ans.
I was my impression at the time because the action followed right after.

[Baldwin]
Q by Counsel
Please state in what manner your tent is enclosed with shade?

[Sieber]
Ans.
Is is very nearly surrounded with shade. There is an opening in front wabout 6 feet wide and facing the north, there is an opening on the East side about 4 or 5 feet wide and facing the East, otherwise the tent is completely surrounded except the little bit in rear and the shade does not quite extend around to the rear end of the tent.

[Baldwin]
Q by Counsel
Did "Kid" get possession of any arms, so for as you know?

[Sieber]
Ans.
No, sir.

[Baldwin]
Q by Counsel
Where were you standing in respect to the door of your tent when the shooting mentioned began that is how far from the door?

Page 100 (3757)
[Sieber]
Ans.
I was standing right at the end of the shade on the right hand side of my ntrance six or seven feet from the tent door.

[Baldwin]
Q by Counsel
As to your opinion that the Indians on horseback looked warlike, did you communicate those impressions to Captain Pierce and were they actually in your mind at the time when you went to the tent?

[Sieber]
Ans.
Yes. My impression was that they looked hostile, I did not say anything to Captain Pierce. I did not have time. As soon as he came there he demanded to know which five were the deserters, and the disarming commenced and gave me no time to tell him my opinion of the fellows outside.

[Baldwin]
Q by Counsel
Could not the cross look and gruff manner of the accused when spoken to by you, have been the result of other causes than any hotile feeling in anticipation of an outbreak?

[Sieber]
Ans.
Yes they might have been caused by some other feelings.

[Baldwin]
Q by Counsel

You say you think two shots were fired at Captain Pierce. Now did you see any one fire them?

[Sieber]
Ans.
I saw two men on the outside whom I concluded must have fired as they had their guns in the position of firing and I saw smoke from their guns.

Page 101 (3758)
[Baldwin]
Q by Counsel
At that moment of time was it not then that the rush by the scouts mentioned by you began?

[Seiber]
Ans.
No. The rush was made a few moments before, Then the firing began.

[Baldwin]
Q by Counsel
You say ou interpreted that Kid gave a look to these four scouts. Now when this look mentioned was given, state where you was standing [where Kid was standing][this phrase was inserted with a carrot in the handwritten transcript.] and where the frou scouts were standing.

[Sieber]
Ans.
Kid was standing about a couple of feet to my right. I was standing back toward my tent like, at the time trying to take his knife and scabbard off his belt. The rest of the four scouts were off to my left.
  The Captain was out a little further and the four other scouts were standing along near together.

[Baldwin]
Q by Counsel
Do you understand the Apache language?

[Sieber]
Ans.
Not thoroughly.

[Baldwin]
Q by Counsel

Do you understand it sufficiently to say whether or not any cry was given by the accused or by other of the four scouts?

[Sieber]
Ans.
There was no cry given by any of the scouts that I heard.

[Baldwin]
Q by Counsel
Did you see any one take any belt or belts?

[Sieber]
Ans.
I saw one of the scouts reaching for his

Page 102 (3759)

Belt at the same time he reached for his gun.

[Baldwin]
Q by Counsel
You have said the accused jumped outside of the shade, now in what direction did he jump, and was it not after shooting began and did you see him again during the supposed occurrences?

[Sieber]
Ans.
He jumped away from me around to the right of the shade. His jumping away from me was almost at the same time as the shooting and that was the last I saw of him during the occurrence.

[Baldwin]
Q by Counsel
You say you thought they had an understanding if a certain thing transpired, which did transpire, that by a look from "kid" each man knew what to do now did all this occur to you at the time, and is not this the result of deductions, conclusions and theories, derived from conversations and after thoughts?

[Sieber]
Ans.
No. My conclusion was formed right there at the time, that the thing was understood.

[Baldwin]

Q by Counsel
When you last saw the accused, did he have any belts or arms in his possession?

[Sieber]
Ans.
No, he did not.

[Baldwin]
Q by Counsel
Was the Scouts so standing that they could see this look, and was this look only a more truning of the head and from what physical movement of face did you interpret the impressions mentioned?

[Sieber]
Ans.
The scouts were standing so that they

Page 103 (3760)
could all see him. When he gave this look he raised his head and looked along the line of scouts who were off to his right.

[Baldwin]
Q by Counsel
In point of time how long did all the occurences prior to the shooting mentioned last?

[Sieber]
Ans.
I don't think it was over five or six minutes.

[Baldwin]
Q by Counsel
You say the accused raised his head and gave a look at that time, what were you doing?

[Sieber]
Ans.
At that time I was taking a knife and scabbard off his belt.

[Baldwin]
Q by Counsel
Could not this look have been other than you say, could it not have been a natural movement or in relation to seeing if all had complied with th e orders?

*[Sieber]
Ans.
No. The look to me was a look from the accused to his men and certainly not accidental.

[Baldwin]
Q by Counsel
How far from the tent door were you when you were wounded?

[Sieber]
Ans.
I was just at the edge of the shade in front of the tent door and six or seven feet away.

[Baldwin]
Q by Counsel
Would you have heard any cry made by the scouts that was made about the time the shooting began?

[Sieber]
Ans.
I could have heard cries, made by the scouts at this time, but I could not have said that it was made by a ascout on some other Indian. Everything

Page 104 (3761)
Was a jumble. I heard a good many cries all at this time.

The Judge Advocate announced that the prosecution here rested.

The Counsel for accused desired to introduce Albert Sieber, already sworn as witness for defense, and testified as follows:

[Baldwin]
Q by Counsel
Prior to these occurrences what has been the general character of the accused?

[Sieber]
Ans.
His general character has always been good.

*The Counsel for accused then asked for a delay until Monday, July 11th as his witness were not here.

The proceedings of the previous day were then read and approved. The court then proceeded to other business before it.

L.D. Tyson
2nd Lieut. 9th Infantry
Judge Advocate
San Carlos, A.T.

8TH DAY

San Carlos, A.T.
July 9, 1887

The court being in session pursuant to the foregoing orders and the counsel for accused having announced that he was now ready with his witnesses the court proceeded with the trial of 1st Sargeant Kid

Present:

1. Major Anson Mills, 10th calvary
2. Captain R.G. Smither, 10th Calvary
3. 1st Lieut. L.P. Hunt, 10th Calvary

Page 105 [3762]

4. 1st Lieut. N DeLany, 9th Infantry
5. 1st Lieut. R.D. Read, 10th Calvary
6. 2nd Lieut. C.P. Johnson, 10th calvary
7. 2nd Lieut. J.B. Hughes, 10th Calvary
8. 2nd Lieut. W.G. Elliott 9th infantry, and
2nd Lieut. L.D. Tyson, 9th Infantry, Judge Advocate,

Counsel and accused present.

Gonshayee, Chief, San Carlos, "I" Band
A witness for the defense their appeared before the Court and having been duly sworn by the Judge Advocate testified as follows

[Baldwin]
Q by Counsel: State your name?

[Gonshayee]
Ans:
Gonshayee.

[Baldwin]
Q by Counsel:
Are you Chief of San Carlos "I" band of Indians?

[Gonshayee]

Yes.

[Baldwin]
Q by Counsel
Were you at Siebers tent on the evening of June 1st 1887.

[Gonshayee]
Ans.
Yes. I was amongst the Indians there at Siebers tent?

[Baldwin]
Q by Counsel
Before going there, or while you were there was there any understanding, argument or arrangement among the Indians, that if this "Kid" was sent to the Guard House that it would be resisted?

[Gonshayee]
Ans.
I don't know anything about it. I don't know whether they made any arrangement about it before they came in or not.

Page 106 [3763]
[Baldwin]
Q by Counsel
Did you see Kid at Sieber's tent?

[Gonshayee]
Ans.
Yes.

[Baldwin]
Q by Counsel:
Was there a disturbance at the tent and if so; did you see Kid go away and in what direction did he go?

[Gonshayee]
Ans.
I saw Kid. He ran toward the Indian Hospital, back of Sieber's tent and back of the Officers tents.

[Baldwin]
Q by Counsel

Did Kid have nay arms with him where you saw him go in the direction of the Indian Hospital, in rear of the line of Officers tents?

[Gonshayee]
Ans.
He had no arms, He had given them up before.

CROSS EXAMINATION

[Judge]
Q by J.A.
Where did you see Kid again after you saw him at the Indian Hospital?

[Gonshayee]
ANs.
I saw him afterwards way up the Gila.

[Judge]
Q by J.A.
Where did you go immediately after the disturbances at Sieber's tent?

[Gonshayee]
Ans.
Of course I heard a whole lot of shots and I went away immediately. I went right between the school buildings in fron of the officers tents.

[Judge]
Q by J.A.
Did Kid get up the Gila before or after you did?

[Gonshayee]
Ans.
We all met together.

[Judge]
Q by J.A.
Was he walking or riding where you came together?

[Gonshayee]
Ans.
He was on foot. I have told you that

PAGE 107 [3764]

I would swear the truth, I will not tell anything that is not the truth.

[Judge]
Q by J.A.
Are you not the Chief of the band to which "Kid" belongs?

[Gonshayee]
Ans.
"Kid" belongs in my band but he has acted as a scout here.

[Court member]
Q by Member
What did "Kid" say after the firing began at Siebers tent?

Counsel for accused objected on following grounds.

[Baldwin]
I object upon the grounds that the question does not relate to the cross examination is in its nature a question for direct examination of the prosecution, and originates evidence by a member of the court.

The court was then cleared and closed, after due deliberation the court was reopened the accused and counsel resuming their seats and the judge advocate announced that the objection is sustained.

> This is only the second time in the trial where one of Baldwin's objections has been sustained.

[Court member]
Q by Court
When you met "Kid" up the Gila, how was it you came to meet him there?

[Gonshayee]
Ans.
Kid got up there on foot, Of course he is better than a horse, can run better than a horse. We were looking for each other. We met coming from various directions.

Saye a witness for the defense then appeared before the court and was objected to by the Judge Advocate on following grounds:

PAGE 108 [3765]
[Judge]
On the ground of interest. The person presented is one of the parties engaged in the outbreak and is incompetent for that reason. The law syas that arties interested are not competent.

108

The counsel replied that
[Baldwin]
the book from which the Judge Advocate takes his view also says that so far as a Court Martial is concerned that objection is not to be considered. (Ives, p. 329) The civil law appertaining to United States Courts makes an Indian a competent witness in trial, hence, if the Judge Advocate objects he must show incompetency. His mere assertion is nothing, he has not shown in a particle of his voluminous testimony that this witness has any interest or knows anything about this case. An accomplice in crime may be a competent witness that only affects its credibility. But as soon as the theory of the prosecution begins to shatter, he objects. If this court wishes to ascertain the truth, I claim this witness should be allowed to testify, all men are presumed to be competent witnesses until the contrary is shown and the assumption of the Judge Advocate can not be taken as evidence of that incompetency.

The court was then cleared and closed, after due deliberation the court wsa reopened the accused and counsel resuming their seats and the Judge Advocate announced that the objection is not sustained.

> Another first – an objection from the Judge Advocate is not sustained.

Saye, a witness for the defense was then duly sworn by the Judge Advocate and

PAGE 109 [3766]
testified as follows:

[Baldwin]
Q by Counsel
What is your name?

[Saye]
Ans.
Saye, San Carlos, A.T.

[Baldwin]
Q by Counsel
Were you at or near Sieber's tent on the evening of June 1st 1887?

[Saye]
Ans.
I was on horseback near Siebers tent that evening.

[Baldwin]
Q by Counsel
Did you see "Kid" there and did you see him go away from there?

[Saye]
Ans.
Yes

[Baldwin]
Q by Counsel:
Were you present at a meeting of five Scouts up the Gila when the subject of coming in and surrendering for an absence without leave was talked about and prior to the time of being at Siebers Tent?

[Saye]
Ans.
I met up the Gila these five Scouts and the same time Mr. Sieber sent word by Gonshayee to come in.

[Baldwin]
Q by Counsel
How many days before you were at Siebers tent?

[Saye]
Ans.
As soon as they got message from Mr. Sieber by Gonshayee they came down the same day.

[Baldwin]
Q by Counsel
Was "Kid" among the member of five Scouts and what did he say up the Gila?

[Saye]
Ans.
He did not say anthing.

[Baldwin]
Q by Counsel
Was there any understanding among the scouts or other Indians that if the scouts were sent to the Guard House that they would resist?

PAGE 110 [3767]
The Judge Advocate objected on the ground that the question is so leading that it almost puts the desired answer in the witnesses mouth. Counsel replied

110

The Counsel admits, that to a certain extent the question is leading, but in view of the fact that it is difficult to get the witness to understand, and who probably can not connect questions and answers in their proper sequence that to prevent a simple question which can be answered be answer by yes or no, forces the defense to confine himself to such a method of questioning that facts cannot be ascertained and the truth shown. The counsel claims that in the interest of all this question should be allowed. Otherwise by reason of the witnesses lack of understanding in respect to the English language, translated into Apache, the defense cannot establish facts.

The Court was cleared and closed, after due deliberation, the court was reopened, the accused and counsel resuming their seats and the Judge Advocate announced that the objection is not sustained.

WOW, the second time his own objection is not sustained.

[Saye]
Ans.
No sir. Those fellows that were with the Scouts came down for the Scouts horses.

[Baldwin]
Q by Counsel
Before coming in, what did the Scouts say that they were going to do?

[Saye]
Ans.
"Kid" said to the scouts, four scouts, We will obey orders whatever Sieber says, they will mind him. If he gave them an order to go to the Guard House they would go, because they went away without permission.

PAGE 111 [3768]
[Baldwin]
Q by Counsel
You say you saw Kid go away from Siebers tent, in what direction did he go?

[Saye]
Ans.
Kid went behind the Officers tents.

[Baldwin]
Q by Counsel
Did he have any arms with him at this time?

[Saye]
Ans.
He had none.

111

CROSS EXAMINATION

[Judge]
Q by J.A.
You stated in answer to the question whether "Kid" said anything up the Gila, that he "Kid" did not say anything now what do you mean by saying afterwards in answer to a question that he "Kid" did say to the Scouts, We will obey orders or whatever Sieber says?

[Saye]
Ans.
So "Kid" did say that, to obey orders as soon as they got down to Sievers tent. Kid did say the scouts must

See next page

PAGE 112 [3770]
surrender. He says he did not understand the question the first time.

[Judge]
Q by J.A.
Did you see "Kid" any more after you saw him going behind the officers tent and where?

[Saye]
Ans.
I saw him afterwards down at the mill.

[Judge]
Q by J.A.
Was "Kid" riding or on foot?

[Saye]
Ans.
On foot.

[Judge]
Q by J.A.
Did he have a horse or mule with him?

Counsel objected on these grounds:

[Baldwin]

112

The question is incompetent on cross examination and does not relate to the direct examination of this witness.

The Judge Advocate replied.

[Judge]
I claim that the Judge Advocate has a right to ascertain all the circumstances of "Kids" getting away in view of the fact that the direct examination has brought in the fact that "Kid" did get away.

The court was then cleared and closed. After due deliberation the court was reopened, the accused and counsel resuming their seats and the Judge Advocate announced that the objection is not sustained.

[Saye]
Ans.
He had not.

[Member of court]
Q by Court:
Are you a brother of the accused

PAGE 113 [3771]
"Kid" now before this court?

The counsel for accused objected upon the ground that
[Baldwin]
the question is irrelevant and incompetent appertains to and has no bearing on this case, an interest in law so far as relationship is concerned is confined to man or wife, and appertains to a challenge to the competency of the witness. More over the question is leading.

The court was then cleared and closed. After due deliberation the court was reopened the accused and counsel resuming their seats, and the Judge Advocate announced that the objection is not sustained.

[Saye]
Ans.
I am not any relation to him.

[Member of court]
Why did you and the other Indians come with the five scouts to Sieber's tent that day?

[Baldwin]
The counsel for the accused objected upon the ground that the question is a question for the prosecution and by a member of this court, and originates evidence for the prosecution.

The court was then cleared and closed. After due deliberation the court was reopened the accused and counsel resuming their seats and the Judge Advocate announced that the objection is not sustained.

PAGE 114 [3772]
[Saye]
Ans.
I came down for my horse and the others came for their horses.

The counsel for the accused then desired to have a delay to close the case.

The court then at 6:20 P.M. adjourned to meet tomorrow on notification from the Judge Advocate.

L.D. Tyson
2nd Lieut. , 9th Infantry
Judge Advocate

9th Day

San Carlos, A.T.
July 11th 1887

The Court met pursuant to adjournment at 8:30 A.M.

Present:

1. Major Anson Mills, 10th calvary
2. Captain R.G. Smither, 10th Calvary
3. 1st Lieut. L.P. Hunt, 10th Calvary
4. 1st Lieut. N DeLany, 9th Infantry
5. 1st Lieut. R.D. Read, 10th calvary
6. 2nd Lieut. C.P. Johnson, 10th calvary
7. 2nd Lieut. J.B. Hughes, 10th Calvary
8. 2nd Lieut. W.G. Elliott 9th infantry, and
2nd Lieut. L.D. Tyson, 9th Infantry, Judge Advocate,

Accused and Counsel present.

The arrear of the proceedings of the two previous days were then read and approved. The accused then presented to the Court four discharges from the U.S. Army. For part of 1882, 1883 and for the year 1884. Signed by Captain Emmit Crawford

PAGE 115 [3773, 3769copy]
3rd Cavalry, as Commanding Officer, and which bears the following character, An excellent Scout upon three and upon one Excellent signed by Lieut. Dugan and Britton Davis, 3rd Cavalry and by Captain Emmitt Crawford, 3rd Cavalry.

The Counsel for the accused there submitted to the Court the discharges and they were found to be as counsel announced.

[Baldwin]
The Counsel requested that the Judge Advocate admit that the defendant has three other discharges of recent years, which bear good character and that owing to the discharges having been worn and torn while in the possession of the accused they cannot be presented to the Court.[25]

[Judge]
The Judge Advocate admitted the above request of the Counsel.

[Baldwin]
The Counsel here stated that as he had no further witnesses to offer the accused desired to make a statement, whereupon the accused made the following statement.

[Kid]
I am 1st Sargeant Kid, San Carlos A.T. I left here without permission from Sieber or the Captain. I went up to Camp and I drank a whole lot of tizwin and got back here in the night, and I passed through here, we went down to kill "Rip" on the Aravaipa. Rip is the man who put up a job to kill m grandfather and "rip said. I shall kill you "Rip says . I have killed several Indians but I never have been tried. As soon as I got up to Rip's camp I saw

PAGE 116 [3774]
"Rip" and shot him, After that I came back, I had been absent five days. I passed my old camp and went up on the foot of the mountain. I was thinking of coming back to Siebers and got back in the evening to Siebers tent. I told all the Scouts they must give up their arms and to obey orders. I was sitting in front of Siebers tent when Sieber came there. Sieber spoke to me, "hello" and I said to him "hello". After that Captain Pierce came. I understood what Captain Pierce said. He spoke to Sieber and asked which of the

---

[25] "Good soldier" evidence is still admissible on the merits if the defendant is on trial for "military" offenses. [Huffman]

Scouts have been absent and Sieber pointed out all of them, and told them to step into line and told me I must give up my belt and gun and also the other four scouts. I laid my gun on ground and belt on the chair. Also the other four scouts put their belts and guns wehre I put mine. Sieber spoke to me "Calaboose". I stepped with my right foot, and Sieber said to me, wait a minute, take off your check from your belt and knife scabbards, after that Antonio Diaz spoke in Apache and said, all the Indians that don't obey the orders, will be sent to Florida. At the same time all the Indians outside made a noise and were much excited about what Antonio Spoke. I thought those outside thought then that we Scouts would be sent down to Florida. At the same time I heard a shot, I was right close to Sieber and facing to Sieber, a chair was standing between him and me, the arms was piled on by the chair and also the belts, Sieber grabbed the guns

PAGE 117 [3775]

And at the same time I skipped out did not pay any attention to getting a gun, and went away in front of Siebers tent south East and heard several shots at that time. Two other scouts besides me, we went in front of Indian Hospital and behind officers tents. I was without arms. I looked back and saw "Tony" shoot at me twice the bullets striking close to me on the ground. I was on foot and went down towards the river. And got up closse to the ditch. I followed up the ditch and passed by the mill, some Indian scouts who were after me shot at me again. I was without arms all the four scouts and some other Indians who had run away I found up on the hill. That is all, after that I went away and I got back here after being absent twenty five days. When I was out in the mountains Gen'l Miles sent word to me that I must come back that IT would be better for me and all my people. I told those four scouts they must all come back and surrender and they did. I had been obeying orders, but God sent bad spirit in my heart I think you all know all the people can't get along very well in the world. There are some good people and some bad people amongst them all. I am not afraid to tell all thse things because I have not done very much harm. I killed only one man whose name is "rip" because he killed my grandfather I am not educated like you and therefore can't say very much. If I made any

PAGE 118 [3776]

arrangement before I came in, I would not have given up my arms at Mr. Sieber's tent. That is all I have to say.[26]

---

[26] The Apache kid made a statement in the case that was apparently unsworn --- something that you would not see or allow an accused to do on the merits of the case in today's proceedings. What I found most amazing was the matter-of-fact discussion of why he had gone AWOL the first time (absent without leave) was to go kill someone named Rip who apparently had killed his grandfather who is the chief of his particular clan. That crime, of course, was not chargeable as an offence under the Articles of War at the time. Today, of course, the Defense Counsel would never allow an accused to testify as to committing a crime like murder as the reason for going AWOL. [Rosen]

The counsel for accused then made the following argument:
[Baldwin]
May it please the court – Gentlemen – We cannot shut our eyes to the fact that in trying this defendant as a soldier of the United States Army we are trying him also as an Indian for the prsecution has studiously labored to connect him, although in becoming a soldier he took the oath of allegiance to our common country, with a band of Indians residing on this reservation and of which band he is supposed to be a popular friend or member. This brings before us two facts. One, the Indian civilization of the nineteenth century, the other the alleged crime of Mutiny, for the alleged crime of desertion fades into insignificance by the side of a crime which when committed might peril the life of an Army, and jeopardize the fate of a nation. At this moment we will review –the charge and its specification in connection with the evidence. Mutiny is the language of the charge. Gentlemen. Bear in mind that the mantle of the law, is thrown around every sccused, upon every stage of the trial and upon every material allegation until torn into tatters by incontrovertible proof establishing guilt beyond all reasonable doubt. In the case of this defendant I do not

PAGE 119 [3777]
Invoke the aid of sentiment or appeal to your sympathies, but upon the evidence I demand for him ou cold hard sense., your conscience and your understanding. Gentlemen of this court. Place side by side the testimony of Captain Pierce Fred Knipple, the so called Chief os Scouts Albert Sieber, and the prophet Antonias Diaz, and I challenge the prosecution to point to a scintillation of proof, that inculpates this defendant in an overact of mutiny. It is true the venerable half Indian, Antonio Diaz saw and heard, what nobody else saw or heard, but it must be remembered that movements impelled by right, create a state of mind and vision, that is capable of hearing and seeing occurrences that never took place. I challenge the Jduge Advocate to establish his beautiful theory of prearranged argument or preconcerted action on the part of this defendant or others from the evidence. That theory vanishes before the light of evidence and of reason, and is but a mirage that might entrap the unthinking. From the evidence, is there ought that inculpates this defendant in a mutiny, and did anyreliable witness see him do one wrong act? Did theyhear him say anything, that indicated insubordinate intentions. I answer for him and then – No. This defendant gave a look

PAGE 120 [3778]
So says the Quartermaster's employee, Siever, and to the mind of Sieber that look said to the scouts present, if I may quote a line of poetry "My chosen braves rush for your guns." Were it not for th solemnity of this trial and its awful gravity, I would say the Quartermaster employee, Al Sieber, the so called chief of scouts has displayed mental powers, that fit him to adorn the chair of psychology in an Eastern College.

Has any witnesses that were at Siever's tent on June 1st 1887 implicated this defendant in a wrong act, on the contrary have they not sworn that this defendant was the first to

117

respond to the orders of Captain Pierce, and give up his arms? Did not all the scouts surrender voluntarily their arms and preparatory to going unforced to the guard house proceed to take their private property from their belts, with he sanction of the Commanding Officer. Did not Sieber assert then in his last proceeding? Was there then a suspicion of hostile intentions there in the minds of those there present? No. Is not much with Mr. Sieber and the prophet Antonio Diaz said, tainted with the suspicion that this defendant came to the tent conscience –

PAGE 121 [3779]

Stricken for previous errors, and with mind formed to do the bidding of lawful authority. Did he not obey? Did he not quietly stand there waiting for final disposition of his case? Has the prosecution shown that htose beyond the limits of the arbor surrounding th tent were there to do his bidding? On the contrary the evidence shown that this defendant, magnificent in physical manhood, fleet of foot and for nigh on to six years the true and trusted scout of our government, stood within the arbor, with head bent and silent fo tongue, perchance trying to hide the shame of his disgrace, by bring degraded full in the sight of his people, and when Captain Pierce spoke and said, look out they are going to shoot" meaning those outside, he raised his head and looked. Twas they the scouts so says Sieber, but Captain Pierce does not say so, rushed for their guns and shots were fired from the outside. Right here comes in the staunch friend of the prosecution the prophet Antonio Diaz who saw and heard what Captain Pierce and Sieber did not see or hear an Indian grab his belt, grab a gun and put a cartridge therein fire a shot in the direction of Captain Pierce and Sieber, and cry out there is nothing for us to do but fight. The testimony

PAGE 122 [3780]

Shows that this valiant son of the latin race Antonio Diaz was within a few feet of Captain Pierce and Sieber, yet they saw nor heard this. The Indian Captive Antonio Diaz's testimony is redolent with the imaginations evolved by the weird incantations of a medicine man. To return to Sieber and "Kid" this defendant Sieber says "Kid" raised his head. That his doing so couldn't have been a natural movement. His look couldn't have been a glance of surprise at the commotion from without and couldn't have been other than what he interpreted it to be. That is if a certain thing transpired which did transpire that look from the stoic face of this warrior among warriors, would say rush my chosen people for your guns. It is passing strange that those sworn to the government who felt uneasy in mind, who were suspicious of trouble, who saw a supposed warlike array, entering the precincts of this post, never uttered a note of warming to the Commanding Officer who was within sound of their voice. Do you remember in this respect the words of the soldier captain Pierce who unconscious of fear, could see nothing in the gathering of the Indians that portended danger. He has lived among them as their agent for two years. So firm in this convictions

PAGE 123 [3781]

that this defendant and the other scouts swould freely obey his orders, without his resort to force of any nature, that he did not even have a guard present. Yet under yon hill was encamped two companies of troops.

Gentlemen of this court. In common reason and in the light of knowledge of Indian character I ask would those scouts have surrendered their arms, if any prearrangement had been entered into, any signal argued upon, or had any design or purpose to evade confinement in the guard house been in their or others words? No. – Say to the door of the Indian every crime that an imbittered press may charge him with, but do not call him a fool. Had this defendant meant war had he afore thought planned a procedure whereby to reek vengeance upon those he had stood by in hours of peril, would he not have led the revolt would not his voice have been heard, would not some movement of hand or arm have unmistakeable ably indentified him with the outbreak, would he have not by aid from those without overmastered the few whites at that tent, and have left minute evidence of his vengeance. But no. He fled unarmed. He vanished from the sight of those at

PAGE 124 [3782]

the tent. Is there any evidence that had Captain Pierce or Sieber have called upon him for assistance that he would not have responded. Had he not given up his arms? Can we say that at the precious moment of time, that his knowledge of our laws and customs enabled him to judge of his duty as a normal soldier. True to the traditions of his race, at the sound of turbulent strife he fled unarmed, why because the history of his people has taught him to scan with suspicious eye the figure of justice, when a white man and an Indian is concerned. It is for you, Gentlemen of this court to judge upon your oaths of the testimony spread only upon the record before this court setting in the trial of this defendant. But I may say it is fortunate for this minfortunate defendant that this court will turn from the atmosphere of prejudice which surrounds the Indian, and which will close its ear to the clamor of those who cry out for the extermination of this people, because they stand in the way of the white mans greed. This defendant a stranger to our language unconscious of our laws, customs, and habits, and spring form a race, still in the darkness of barbarism, and who burns to ashes the

PAGE 125 [3783]

Habitations, and the body of those who die and scatter to the north winds the ashes of the dead, cannot reasonably be viewed as guilty of wrong by mere and unavoidable presence of an outbreak which spontaneously originated, without assistance aid or encouragement from him. The testimony of our Duchin is unimportant to the mind of the counsel that the almost forgot to mention it. Here let me remark. And the Judge Advocate may correct me, if I misquote, that Wharton on Criminal Evidence lays down the rule that testimony of prejudiced witnesses is tainted with discredit and is valueless in comparison with that of unbiased witnesses. Now the reply of this worthy blacksmith Mr. Duchin to the question as to his bias and prejudice manifestly indicates prejudice against this defendant either as individual or against him on account of his race. In

119

either event the bias taints his testimony, particularly where he swears to the accused having fired a shot from a point, between an outhouse and the dormitory, which pleace gentlemen you know was in sight of many who witnessed the occurrences

PAGE 126 [3784]
Of June 1st, 1887, and not one of whom has been called to corroborate the testimony of the knight of the anvil, on Duchin. Moreover gentlemen the occurrences of June 1st 1887, glaring and full in the sight of the community, (I quote from the record) would have occasioned some one to have observed an Indian as prominent around this post as the first Sargeant of the indian Scouts. Yet none did so.
Although the place from whence the shot mentioned by Duchin as fired was right in the front and view of the line of officer's tents. The testimony of the witness who relates that his suspicions were arousned because the Indians were uncommunicative and not disposed to talk, and that they evidenced their utterance by saying "No Sabe" needs no review. You may not believe the evidence of the Indians who gave testimony if you rely upon the opinions of the Judge Advocate, but let me call your attention, to the fact that they have been unimpeached on the contrary they have been corroborated. The all seeing eye of Him who knows men's heart has recorded on the book of Judgement each the truth of their assertions

PAGE 127 [3785]
I ask you to remember for we have no diagram of the places or buildings mentioned in the testimony that the direction taken by this defendant could in no way bring him near the place or buildings that the prejudiced mind of Duchin put him. Moreover the testimony of the Indians corroborated by Sieber and Captain Pierce shows that the accused was unarmed, and when seen again up the Gila and at the mill he was on foot. Gentlemen of the court. In Law there is charity as well as rigor. The iron grasp of government does not seek to crush the ignorant when that ignorance is the outgrowth of ages. The enlightened principles of humanity seeks to first elevate to the standard and plans of civilization , which centuries of intelligence has given us as white men, before it condemns the rude barbarian by our merciless law and inflexible articles of war. In crime and particularly the crime of Mutiny, there is an essential Element. Intent, hence from the evidence it is affirmatively shown that this defendant was only present at the outbreak , he aided not, he assisted not, he countenanced not, he abetted not., but quick to foear the white man and when no calling voice for him to stand by his duty as a

PAGE 128 [3786]
Soldier was heard, he fled unarmed. Suspicious and prejudices that have become a part of his nature were more powerfull to impel flight than a knowledge of intricate laws which he never heard nor could not comprehend. Would be for ehim to stay and defend his Commanding Officer. Mark the working of the specification. Is there any evidence that this defendant resisted arrest, seized his arms or fired upon his Commanding Officer, If there is no evidence against him is he not innocent? I cannot believe the

prosecution is so eager to wear a virgin wreath of victory, that it will demand that this defendant shall prove himself innocent before testimony connects him directly with crime. It is true that the testimony contradictory and saturated with the conclusions and deductions of vivid imaginations, creates an inference that others fired upon the Commanding Officer, but has there been shown any connection between this defendant and them? On the contrary has not the prosecution signally failed? Did ont this defendant do what almost any one would have done ? Leve the scene of disorder and particularly when that one is not versed in intricate and complex questions of law and duty. Gentlemen of this court. This defendant is a soldier yet his childlike faith in the

PAGE 129 [3787]

superior wisdom of those controlling his destinies caused him to submit to the deprivation of many of his rights as a soldier. Though a soldier in name, yet in nature and intelligence he is an Indian, and as counsel, I set up the claim, that the broad principle of human justice, that only condemns the actually guilty upon which law in its technical sense is founded debars legal maxims from consideration in this case and excludes construction crimes as not within the view of the law, when ascertaining the actually guilty. You are now sitting in judgement upon the Indian Civilization of this century. , but that is not for me to discuss, but I ask you to weigh the evidence, and view this defendant in the light of the progress of that civilization. If this defendant has done wrong he waits your verdict. He came here voluntarily, he can not see, either by the light of the evidence or the promptings of conscience where he has erred beyond an absence. He fears no punishment that right and justice will decree. From the silent gorges and rocky fastnesses of the serriea Mountains that dot the face of this desolate country, at the bidding of General Miles this defendant came back and surrendered. For this

PAGE 130 [3788]

unhappy defendant, I ask you upon the evidence that when weighed in the even scale of Justice, and when viewed with mature deliberation, that you do not condemn this accused.

To the Judge Advocate I extend my thanks for his courtesy, and the fairness with which he has performed his duties. To the Court I return my thanks for its patience kind consideration and courteous attention.

The Judge Advocate submitted the case without remark.

The court was then cleared for deliberation and having maturely considered the evidence adduced finds the accused First Sargeant "Kid Co. A" Indian Scouts:

Of the First Specification, First Charge.

"Guilty" with the exception of the word "his" before the word "arms" and of the excepted word, "Not Guilty".

Of the First Charge:
"Guilty"

Of the First Specification, Second Charge:
"Guilty"

Of the Second Charge
"Guilty"

PAGE 131 [3790]
And the court does therefore sentence him, First Sargeant "Kid," Co. A, Indian Scouts, to suffer death by shooting with musketry, two thirds of the members of the Court having voted for this sentence.[27]

The court then at 12 N. adjourned to meet again at 1:30 P.M.

Anson Mills
Major 10th Cavalry
President

L.D. Tyson
2nd Lieut., 9th Infantry
Judge Advocate

---

[27] I think the defense counsel did a good job. I think the cards were stacked against the Kid in this case. Clearly you had members that had expressed an opinion about the case or formed an opinion about his guilt. Most of the members indicated that they were biased against the Kid so I think the conclusion, the findings and sentence of the court were a foregone conclusion. [Rosen]

Contributions to what we know about Apache Kid

The transcript revealed some details about the Apache Kid that have been unclear in the historical academic literature. For example, when the Apache Kid began his service as an Army Scout and who recruited him have been unclear.[28]

The Apache name for Apache Kid was given in seven different ways by different people,[29] During the trial, one of the witnesses for Apache Kid said his name was "Hahouantell". Sayes, another Apache Scout, said his name was "Shisininty".[30] He was known as "the Kid" but later became known as "Apache Kid" and to be consistent, this is the name that is used throughout this book.

From the transcript we learn that Capt. P.L. Lee, 10th Cavalry, enlisted the Apache Kid, April 11, 1887 at San Carlos, A.T.. The court report wrote 1887, but this was obviously a mistake. We may never know what Capt. Pierce stated as his enlistment year, but from the discharges, he was enlisted in 1879 or 1880. The defense counsel, Lt. Baldwin, presents discharges for the Apache Kid, beginning in 1880:

> I would like to proffer for discharges from the US Army for the period of 1880 to 1883 and for year 1884 signed by Captain Emmett Crawford cavalry is commanding officer and which bears the following character an excellent Scout upon three and upon one excellent signed by lieutenant Dugan and Britton Davis 3rd cavalry and by captain Crawford 3rd cavalry I also respectfully request the judge advocate admit that the defendant has three other discharges of recent years with very good character and that owing to the discharges having been worn in for and while in the possession of the accused they cannot be presented to the court. [31]

[Judge Advocate]
Was he a duly enlisted Soldier?
[Capt. Pierce]

---

[28] Clare McKanna, Jr., Renegade of Renegades 13 (2012) writes with no source that Apache Kid enlisted in 1882.

[29] Radio interview of Clare McKanna, Jr.

[30] Clare McKenna, Jr., *Renegade of Renegades* 6 (2012).

[31] General Court Martial of Apache Kid, Company A, Indian Scouts," Records Received, Transcript, p. 115, December 27, 1887, Judge Advocate General (Army), RG 153, National Archives.

Answer: He was. He was enlisted by Capt. P.L. Lee, 10th Cavalry on the 11th day of April 1887 at San Carlos A.T.
Question by J.A.:
Do you know if the oath was duly interpreted to him and by whom.
[Capt. Pierce]
Answer: I do. I don't remember if it was by McIntosh or Antonio Diaz. I think by Antonio Diaz.[32]

In testimony from Capt. Pierce, the date that Apache Kid wanted to return was actually May 29. The messaging continued until he arrived on June 1 in the evening around 5pm which the alleged incident of mutiny occurred.

> *Capt. Pierce*
> *Answer: I saw him about sundown in front of Mr. Sieber's tents. He absented himself from camp without permission on the evening of the 29th or morning of the 28th of May 1887. On the 30th of May the messenger alleging that he came from Sargeant Kid, said that he "Kid" wished to see me, I replied that he could see me if he wished to but that I should make no promises whatever. On the morning of the 1st of June the messenger "gonshayee" came again and said that he came from Kid and that "Kid" wished to see me. I told him that he could come if he pleased and that the sooner he came the better it would be for him. About sundown on the first of June I was informed that he had returned and that he was at Mr. Sieber's quarters. I went there with Mr. Seiber and also with Antonio Diaz.[33]*

In testimony from Capt. Pierce, the date that Apache Kid had been absent puts the date at May 25, 1887 when he left to kill Rip. Then he came to turn himself in on June 1, 2017. The second absence was from June 1 until June 25. This has been reported occuring two weeks after his Father/Grandfather was killed.[34]

Question by Judge Advocate
How many days had the accused been absent without authority?

[Capt. Pierce]
Ans:
Thirty days. Five days the first time, and twenty five days including the first of June afterwards.[35]

---

[32] General Court Martial of Apache Kid, Company A, Indian Scouts," Records Received,Transcript, p. 35, December 27, 1887, Judge Advocate General (Army), RG 153, National Archives.

[33] General Court Martial of Apache Kid, Company A, Indian Scouts," Records Received, Transcript, page 43, December 27, 1887, Judge Advocate General (Army), RG 153, National Archives.

[34] Paul Andrew Hutton, The Apache Wars, 393 (2016).

[35] General Court Martial of Apache Kid, Company A, Indian Scouts," Records Received, Transcript, p. 44, December 27, 1887, Judge Advocate General (Army), RG 153, National Archives.

One author suggests that it was Gonshayee who persuaded him to carry out his obligation to kill Rip.[36]

According to testimony from Capt. Pierce, Apache Kid left with thirteen men to kill Rip, with only six from his band staying back:

[Pierce]
Ans:
I think he has considerable influence in his band. Thirteen men of his band went with him, while six stayed at home and did not go.[37]

---

[36] Paul Andrew Hutton, The Apache Wars, 393 (2016).

[37] General Court Martial of Apache Kid, Company A, Indian Scouts," Records Received, Transcript, p. 44, December 27, 1887, Judge Advocate General (Army), RG 153, National Archives.

Postscript

The story of the Apache Kid and his ability to elude capture for the rest of his life has taken its place in western legends in the United States. Within his community at San Carlos Apache Reservation, today, he is respected among his family descendants and remembered in song among the San Carlos Apache, particularly in the Seven Mile region of the San Carlos Apache Reservation. He is remembered an outstanding U.S. Army Indian Scout, with exceptional skills and talents. He carried his honorable discharges with him until they were "worn and torn" according to the transcript.[38]

Following the trial, the convening authority General Miles in San Francisco, California, determined the sentence of the death penalty was too harsh. On July 29, 1887, he wrote a letter to Gen. Anson Miles disallowing the death penalty and ordering the court to reconvene and reconsider the sentence. It had been General Miles who had contacted Apache Kid, urging him to return to the San Carlos Army Post after being away without permission. He may have thought that Apache Kid, having returned at his urging, should not be rewarded with the death penalty.

General Miles ordered the court to reconvene at another Fort, and reconsider the sentence. They reconvened at Ft. Smith about 30 miles away, on August 6, 1887 and returned a sentence "to be dishonorably discharged the service of the United States forfeiting all pay and allowances and to be confined at hard labor as such place as the receiving authority may direct for the period of his natural life."[39] General Miles had the authority to further reduce the sentence, which he did, on December 27, 1887, writing, "The period of confinement is mitigated to ten 10 years. The sentence mitigated is onfirmed and will be duly executed. The military prison at Fort Leavenworth Kansas, is designated as the place for the execution of so much of the sentence as relates to confinement of hard labor where the prisoner will be sent upon receipt of further orders from these headquarters."[40] General Miles then sent the sentencing recommendation to the Adjutant General, Lt. Gen. Sheridan. On January 23, 1888, Special Orders No. 18 reassigned Kid

---

[38] "General Court Martial of Apache Kid, Company A, Indian Scouts," Records Received, p. 115, Judge Advocate General (Army), RG 153, National Archives.

[39] "General Court Martial of Apache Kid, Company A, Indian Scouts," Records Received, Misc. Correspondence, December 27, 1887, Judge Advocate General (Army), RG 153, National Archives.

[40] Ibid. Letter from Gen. Miles. December 27, 1887.

from Leavenworth to Alcatraz.[41] The Acting Judge Advocate General. J. Norman Lieber in Washington, D.C. saw bias and other flaws in the evidence and charges from the record. April 11, 1888, he wrote a 45-page appeal to the Secretary of War, asking that he remit the sentence of Apache Kid and that of the other four scouts. He wrote his appeal for all five of the scouts who had originally all been given the sentence "to suffer death by shooting with musketry."[42] Judge Advocate Lieber begins by reviewing the biases in the voir dire questioning of the members of the court. He writes:

> Considering their testimony elicited in the examination of these officers as to their competency as members of the Court, together with the telegraphic statement of the Court to the convening authority dated June 29, 1887, it is evident that the numbers challenged, although disavowing any bias or prejudice against the prisoners personally by reason of having no knowledge of their connection with the affair of June 1, 1887, had formed an opinion as to the guilt of the parties, there unknown to them, which opinion must have become operative the moment the prisoners were shown to have been connected with the occurrence under investigation. This preconceived opinion was not consistent with the rights of the prisoners to a fair and impartial trial.[43]

> . . . it is deemed necessary , before proceeding to the consideration of the evidence to examine the preliminary proceedings of the court in the remaining cases.[44]

The Judge Advocate quoted the language from the letter from Gen. Miles in his appeal:

> While the prisoners are enlisted scouts, the evidence shows them to be ignorant unlettered Indians, requiring even the rendering of the language of the court through two languages or mediums to convey to them an idea of what was said and done before the court in their cases not wearing the uniform of the service; acting under the direction of a civilian chief of scouts; uninstructed in the code of crimes and punishments prescribed by the Articles of War; unfamiliar with the discipline of the service and uninformed as to the responsibilities

---

[41] General Court Martial of Apache Kid, Company A, Indian Scouts," Records Received, Misc. Correspondence, Jan 23, 1888, Judge Advocate General (Army), RG 153, National Archives.

[42] General Court Martial of Apache Kid, Company A, Indian Scouts," Records Received, Misc. Correspondence, Letter of Appeal from Gen. G. Lieber to Sec. Endicott, Apr. 11, 1888, p.2, Judge Advocate General (Army), RG 153, National Archives.

[43] General Court Martial of Apache Kid, Company A, Indian Scouts," Records Received, Misc. Correspondence, Letter of Appeal from Gen. G. Lieber to Sec. Endicott, Apr. 11, 1888, pp. 2-26, Judge Advocate General (Army), RG 153, National Archives.

[44] Ibid.

of a soldier; not under the same rigid restraint but mingling freely with their band, sharing it sympathies and participating in its customs and ceremonies.[45]

It appears the prisoners accompanied their band to execute the system of summary punishment practiced among them and recognized by Section 2146 of the Revised Statutes of the United States. For this the authorities at the Agency were dissatisfied and determined upon their arrest. . . . Upon their return, the Agent proceeded to disarm and confine them, but without an expressed understanding as to the particular nature of their offending when Antonio Diaz, a Mexican interpreter of their language present, announced or gave them to understand they would be banished to Florida or the Islands for not obeying orders, or something to that effect.

Each of the prisoners asserts this in his statement before the Court, and they are coorborated that such an announcement was made by the evidence of their chief (who was present at the time). The fact that such announcement was made, was not questioned until the close of the trial of the 5th case, which Antonio was recalled and he denied that he made it. But it seems incredible that these prisoners who have heretofore borne good characters for reliability, obedience and friendship for the white race, should suddenly become mutinous, break away and escape from the reservation without some serious grievance, real or fancied, impelling them to the act. It is highly probable that the statement about banishment was made at this time; and immediately thereupon other members of the band loaded their guns and opened fire on the tent where Captain Pierce and the chief of scouts were receiving the arms of the prisons and then ran away. . .[46]

The disarming and ordering into confinement and the announcement of banishment – simultaneous acts without explanation to disabuse the minds of the prisoners of their erroneous impressions, resolved them to fly at all hazards rather than to submit to banishment for what they believed was an unjust infringement upon their rights.[47]

Further, the Judge Advocate Lieber made the observation that the testimony showed only two guns were missing and there were five prisoners charged with firing guns. He concludes with: "Reviewing the trials as not having been fairness and the evidence not sustaining the findings, it only remains to recommend that the sentences be remitted."[48]

The Secretary of War, with a clear dislike of Apaches, wrote, "As these Indians have been in confinement such a short time, the Secretary of War will take no action at present but desires to

---

[45] General Court Martial of Apache Kid, Company A, Indian Scouts," Records Received, Misc. Correspondence, Letter of Appeal from Gen. G. Lieber to Sec. Endicott, Apr. 11, 1888, pp. 43-44, Judge Advocate General (Army), RG 153, National Archives.

[46] General Court Martial of Apache Kid, Company A, Indian Scouts," Records Received, Misc. Correspondence, Letter of Appeal from Gen. G. Lieber to Sec. Endicott, Apr. 11, 1888, pp. 42-45, Judge Advocate General (Army), RG 153, National Archives.

[47] Ibid, p. 46-47.

[48] Ibid., p. 47.

have the matter laid before him again six months hence. Chief Clerk." On October 11, 1888 Gen. Lieber resubmitted the appeal and it was granted, as promised, by Sec. Endicott, and he remitted the sentences of the Apache Kid and his four scouts. On October 29, 1888, the Commanding Officer, O.O. Howard, sent a letter to release the prisoners from Alcatraz and transport them back to San Carlos.[49]

Historians write that he traveled by train to Casa Grande and then took the stagecoach to San Carlos.[50] from accounts of the blacksmith that he was greeted by the 10th Calvary Band playing for him and welcoming him back to the U.S. Army post in San Carlos.[51]

The Federal Criminal System

When Apache Kid arrived back in San Carlos, not everyone was pleased. Within within days of his return he was arrested on a murder charge, because the local community was unhappy about his release from prison.

Captain Bullis, the Federal Indian Agent, filed a complaint in federal court against Apache Kid and others for the murder of Michael Grace near Crittenden, Arizona. He was arrested November 17, 1888 and transported to Tucson, to the Pima County Jail. Murder charges were not presented against him until March 4, 1889, but he was held in custody until the grand jury hearing.[52]

Meanwhile, the Washington, D.C.-based, civil rights group, the Indian Rights Association, brought the matter of federal jurisdiction over Indians before the U.S. Supreme Court.[53] Gonshayee, the chief of Apache Kid's band was the lead plaintiff in the habeaus corpus action to free him from his confinement for his conviction of murder of William Diehl [54]on June 5, 1887 in a federal circuit court in Arizona. The federal court had sentenced them to death by hanging: *Friday, the 10th day of August, A.D. 1888, and on that day you be taken by the United States marshal of the Territory of Arizona, to and within the yard of the jail of said Maricopa County, Arizona, and between the hours of nine o'clock A.M. and five o'clock P.M. of that day, by said marshal, you be hanged by the neck till you are dead.*[55]

---

[49] General Court Martial of Apache Kid, Company A, Indian Scouts," Records Received, Misc. Correspondence, October 29, 1888, Judge Advocate General (Army), RG 153, National Archives.

[50] Paul Andrew Hutton, The Apache Wars 398 (2016).

[51] Phyllis de la Garza, The Apache Kid, pp. 64-64 (According to the Army Blacksmith, Edward Arhelger, he wrote that the scouts were "serenaded" by the 10th Calvary band as they entered camp.

[52] Clare V. McKanna, Jr., Renegade of Renegades 139-140 (2016).

[53] Paul Andrew Hutton, The Apache Wars 400 (2016).

[54] *The Arizona Champion*, June 18, 1887.

[55] *Ex parte Gonshayee*, 130 U.S. at 346-347 (1888).

The opinion documents that the Indians had no defense to the crime of murder.[56] However, their defense counsel, Alexander and Chalmer, requested payment from the Treasury of $2000[57] for their representation of "fifteen Indians" in a murder case. The amount had been approved by the judge, according to the document. The U.S. Department of Justice had no appropriations budget for court-appointed counsel, according to the letter and Congress was asked to pay for their representation.[58]

> The request for payment included the next steps for the defendants:
> Tho first ten cases are appealed and must be prepared for the Supreme court. The two cases on writ of habeas corpus are prepared for the Supreme Court of the United States and are to be briefed; the attorneys claim that a reasonable fee for all the work is $2,000, and that the Government should pay for the defending these people. We thus make our claim.
> H. N. ALEXANDER.
> L. H. CHALMERS.
> I do hereby certify that the above services were performed. Judge Alexander proposes to present two of the cases to the United States Supreme Court, by habeas corpus; one on jurisdiction of the court for crime committed without, and for crime committed within the reservation.
> I deem it of importance that the act of March, 1885, should be acted upon and finally settled by United States Supreme Court. There is great diversity of opinion by tbe bar of the Territory, and in view of this I think the above fee of $2,000 is reasonable.
>
> WM, W. PORTER,
> District Judge.[59]

The Indians whose appeals went forwarded were heard in the May 29, 1888 term, the U.S. Supreme Court issued its opinion in *Ex parte Gonshayee*[60] explaining that criminal complaints, not federal crimes, against Native Americans must be adjudicated in the state and territorial courts, and federal courts had no criminal jurisdiction over individual Indians for territorial crimes.

---

[56] Ex parte Gonshayee, 130 U.S. at 346 (1888) *The defendant was then asked if he had any legal cause to show why judgment should not be pronounced against him; and no sufficient cause being shown or appearing to the court, thereupon the court renders its judgment that, whereas you, Gon-shay-ee, having been duly convicted in this court of the crime of murder, it is found by the court that you are so guilty of said crime.*

[57] According to the inflation calculator, $2000 in 1888 is equivalent to $50,811.17 in 2018. See https://www.officialdata.org/ (last visited May 29, 2018).

[58] H.R. Exec. Doc. No. 381, 50th Cong., 1st Sess. (1888).

[59] H.R. Exec. Doc. No. 381, 50th Cong., 1st Sess. (1888).

[60] *Ex parte Gonshayee*, 130 U.S. 343-53 (1889).

The time from the conviction in the federal court and the decision of the U.S. Supreme Court before carrying out the death sentence, was June 4, 1888 to April 15, 1889. The Arizona Champion, August 4, 1888 reported on Gonshayee being held in Phoenix in a dehumanizing observation about his state of uncertainty. Procedurally, these cases are complex, and the western legal concept of due process could not have squared with the Apache sense of due process as in the act of retribution killing that settled all murder cases.

Phœnix Arizonan: Gon Shay-ee, the Apache Indian sentenced to be hung on the 10th, is evidently not aware of his new lease of life, secured through a stay of proceedings pending an appeal to the Supreme Court, as he is very anxious to have his family, at San Carlos, come and see him before he dies. He has a photograph of his two wives and three children, which he holds in his hands almost the entire time.

Aug 4, 1888, The Arizona Champion

Return of Indian Murderers.

Twelve Apache Indians, convicted of virious crimes, principally murder and sentenced to different terms imprisonment, and who have been discharged from prison under the recent decision of the United States Supreme Court holding that Indians should be tries in the Territorial and not the United States Court arrived here Thursday night in charge of Lieutenant B. C. Lockwood of Columbus baracks, Ohio, and two Infantry Sargents, and yesteday left for San Carlos in company with a number of thier compatriots who had come here to receive them. Eleven of these Indians came from the Ohio state prison at Columbus. and one from the State prison at Springfield, Illinois. They were all dressed in American style, a la mode, and presented quite a natty appearance. They are all hard characters, however, and thier release from prison and returned to San Carlos Reservation can be regarded in other light than a public calamity.— Wilcox. Stockman.

June 11, 1889 The Arizona Champion

Based on the Supreme Court decision in Ex parte Gonshayee, Apache Kid, along with any other Indian held or convicted by a federal court for a state or territorial crime was released. The U.S. Supreme Court opinion allowed for the cases to be brought in the territorial court.

The Territorial Criminal System

Gonshayee's murder trial

Their release was short-lived because the case was immediately filed in the territorial court for murder and Gonshayee, Nahconquisay and Akisaylala were each tried for murdering William Diehl in their 1887 raid.

This article demonstrates the dehumanizing context in which the Apache Scouts had to defend themselves. The political connection of Mike Grace was also noted in the same newspaper.

> FLORENCE will soon have a hanging picnic of five Apaches—three for the murder of William Deihl, and two for the murder of William Jones. A step in the right direction. If the civil authorities had delt with these "Red Devils" all along since Arizona became a Territory, the Apache troubles would have been quelled long before they were, and the millions which have been expended to sublue them by military, would have been saved to the country.—Florence Alta.

Oct. 26, 1889, The Arizona Champion

> THE recent escape of a number of Indians from the San Carlos reservation and the subsequent murders of William Diehl and Mike Grace show what a treacherous set of red devils they are and how impossible it is to prevent them from committing such crimes. The whole crew must be fired out of the territory before human life will be secure. If the Indian Friendly Bureau object to their imprisonment, ship the whole gang out to their care, when they can place them in a New England dime museum and pet them to their hearts content.

June 18, 1887, The Arizona Champion

> Mike Grace, who was killed by the Apaches is said to be a cousin of ex-Mayor Grace, of New York City.

June 25, 1887, The Arizona Champion

Gonshayee, Nahconquisay and Akisaylala were sentenced in to be hanged in Florence, Arizona.[61] Before their sentence could be carried out, the three men, in a suicide pact committed suicide.[62]

---

[61] Clare V. McKanna, Jr., White Justice in Arizona: Apache Murder Trials in the Nineteenth Century, 92-93, 178-79 (2005).

[62] Paul Andrew Hutton, The Apache Wars 400 (2016).

Apache Kid's murder trial

Despite the fact that Apache Kid had already been tried for the mutiny claim that he shot or was responsible for shooting Al Sieber, this was a criminal complaint of attempted murder for that same incident. Al Sieber clearly wanted to convict Apache Kid, although it was clear that he knew Apache Kid had not shot him. Al Sieber was the main witness against him, supported by Capt. Pierce and Capt. Bullis.

The Sheriff, Glenn Reynolds, called on Gen. Miles for support from the Army in arresting Apache Kid for the crime. Gen. Miles was clearly aware of the facts in that case, but agreed to assist the Sheriff in his capture and assigned Federal Indian Agent, Capt. Bullis to the task.[63]

The Constitutional Fifth Amendment protection against double-jeopardy applies in the same jurisdiction, but the military, the federal government and the territorial government were all different jurisdictions.

If Apache Kid and the Apache people were told or advised of the civil rights available to any person in the jurisdiction of the United States, including that of double jeopardy clause, trying the same incident, yet a different crime, would have been highly confusing. Gila County District Attorney, J.D. McCabe issued a warrant for his arrest was issued on October 14, 1889.

Apache Kid along with the same three defendants as in the court martial case, were tried in the Gila County Courthouse on October 29, 1889. Judge Joseph H. Kibbey, presided and two court-appointed attorneys represented Apache Kid and his co-defendants. The court interpreter was someone familiar to Apache Kid, Merejildo Grijalva.

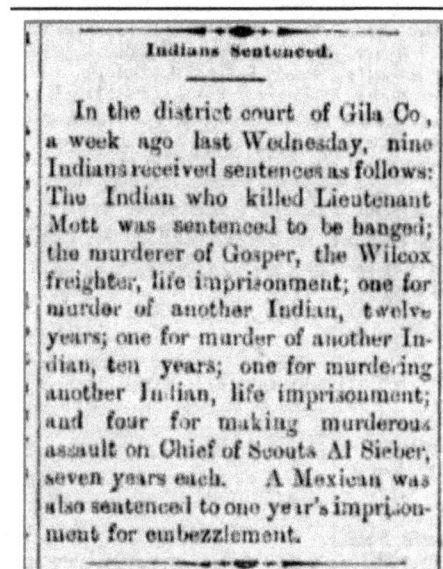

Indians Sentenced.

In the district court of Gila Co., a week ago last Wednesday, nine Indians received sentences as follows: The Indian who killed Lieutenant Mott was sentenced to be hanged; the murderer of Gosper, the Wilcox freighter, life imprisonment; one for murder of another Indian, twelve years; one for murder of another Indian, ten years; one for murdering another Indian, life imprisonment; and four for making murderous assault on Chief of Scouts Al Sieber, seven years each. A Mexican was also sentenced to one year's imprisonment for embezzlement.

Nov 9, 1889, The Arizona Champion

---

63 Paul Andrew Hutton, The Apache Wars 400 (2016).

The events of the trial were told by the blacksmith, Ed Arhelger, who gave his perspective on the trial.[64] Ed Arhelger sat in the chair in the courtroom to hear the case. According to his account, Al Sieber testified that Kid shot him, even though he knew that he did not even have a weapon in his hand during the incident. Antonio Diaz and Frank Porter both supported Sieber's testimony.[65] Vacasheviejo and Fred Knipple testified against Apache Kid and the other defendants.[66] Following their testimony, a scout named Curley, identified Apache Kid as the shooter. Ed Arhelger had been at the incident and saw Curley shoot Al Sieber, but Ed Arhelgerc was not called as witness for Apache Kid.

Apache Kid testified. He said that Curley had been a rival for his wife, Nasahasay, known as Beauty, and testified Curley was bitter because she rejected him, and became Apache Kid's wife. In addition to that, Apache Kid said that Curley was resentful because Al Sieber preferred him as a scout. Kid stated that he would never shoot Al Sieber. Beauty, Apache Kid's wife, took the stand and testified to his character and supported his story about Curley's jealousy.

The prosecutor recalled Al Sieber to the stand and Sieber testified that Apache Kid made up the whole story about Curley.

Both parties gave closing statements and the jury returned a verdict of guilty for Apache Kid and his three accomplices. They were sentenced the next day by Judge Kibbey to seven years in the Yuma Territorial Prison.[67]

————————————————————

Sheriff Glenn Reynolds was given the task of transporting some prisoners, including Apache Kid, to the Yuma Territorial Prison to begin their sentences. The conditions of the prison was deplorable, and statistics show that Indian death rates were 37% in that prison. On the journey, some of the prisoners who were Apache scouts turned on their captors and managed to free Apache Kid and they escaped into the mountains.

Apache Kid was never recaptured. The Army offered a reward of $5,000 (equivalent to about $129,518.68 in 2018).[68] There is a consensus that he went to Mexico. Sightings continued for years, and more than seven people claimed to have killed him. In 1907, Yale University's Skull and Bones Society claimed to have his skull from one of these claimants, continuing dehumanizing Apaches.

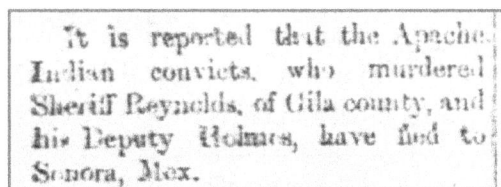

"It is reported that the Apache Indian convicts, who murdered Sheriff Reynolds, of Gila county, and his Deputy Holmes, have fled to Sonora, Mex.

November 16, 1889, The Arizona Champion

---

[64] Paul Andrew Hutton, *The Apache Wars* 401 (2016).

[65] Paul Andrew Hutton, The Apache Wars 401 (2016).

[66] Clare V. McKanna, Jr., Renegade of Renegades 141 (2012).

[67] Paul Andrew Hutton, *The Apache Wars* 401 (2016).

[68] https://www.officialdata.org (last visited May 29, 2018).

What became on the members of the court and the witnesses?

Albert Sieber, Chief of Scouts may be the most well-known of all the participants in the trials of Apache Kid. He was a German immigrant and was a civilian miner before being persuaded to join as Chief of Scouts in Arizona. The transcript does away with the previous accounts that suggest he was sympathetic to the Apaches and left because he thought they were being mistreated, when he was the impetus behind trying to give Apache Kid and his Scouts the death penalty and did not give up until he was convicted for attempted murder on himself in the incident he claimed was a mutiny attempt in a territorial court.

Al Sieber left his job as Chief of Scouts at San Carlos reportedly because of a conflict with Capt. Bullis, who fired him Dec. 1890.[69] He died when a boulder rolled on him while he was supervising a crew in the construction of the Tonto Dam near Globe, now the Roosevelt Dam.[70]

2d Lt. L.D. Tyson, Judge Advocate in the court martial, moved back to Tennessee and became a successful textile magnate. He earned a law degree from the University of Tennessee and practiced law. He was on the ballot for Vice President at the 1920 Democrat National Convention, but did not win the nomination. In 1925, he was elected to the U.S. Senate. He passed away while still serving in the Senate from unknown causes. His great-granddaughter is the 28th President of Harvard University (2007).[71]

Gen. Anson Mills, President of the Court, was promoted to Brigadier Gen. 1897 and retired. He started an ammunition belt business, the Mills Woven Cartridge Belt Company, of Worcester, Massachusetts, which manufactured woven cartridge belts and equipment for all the world and supplied the Spanish American War. He made a small fortune by 1905.[72]

1st Lt. John Arthur Baldwin, 9th Infantry was West Point graduate and a graduate of the infantry and Cavalry School in 1885. He was appointed from New York, Second Lieutenant, 9th U. S. Infantry, 27 July 1872, an African American unit, except for the officers. He became First Lieutenant, 19 May 1881; Captain, 4 November 1890; Major,[73] 22nd U. S. Infantry, 2 June 1899 ; Lieutenant Colonel and Colonel, 16th U. S. Infantry, June 1902. He was sent to the Philippines,

---

[69] Paul Andrew Hutton, *The Apache Wars* 409 (2016).

[70] Paul Andrew Hutton, The Apache Wars 423 (2016).

[71] https://en.wikipedia.org/wiki/Lawrence_Tyson .

[72] Gen. Anson Mills Arlington Cemetery database at http://www.arlingtoncemetery.net/a-mills.htm (last visited May 28, 2018).

[73] Edgar C. Emerson, ed., "Our County and it's people: A descriptive work on Jefferson County, New York, The Boston History Co., Publishers (1898) at http://www.onlinebiographies.info/ny/jeff/baldwin-ja.htm (last visited May 28, 2018).

leading an African-American unit, wrongly determined to be immune to tropical diseases. He died 15 March 1903 from a tropical disease from his service in the Philippines. He is buried in Arlington National Cemetery.[74] His son, followed in his footsteps, graduating from West Point and becoming a decorated World War I General.[75]

2d Lt. Carter Johnson 10th Cavalry. C.P. Johnson was described as a "character" by Harper's Weekly, where he is pictured in a spread of photos, in an article written by Federic Remington.

In the Amon Carter Museum of Fine Art, the following paintings by Frederic Remington describe their subject as 1ˢᵗ Lt. Carter Johnson.

June-July
Remington travels to New Mexico and Arizona for The Century Magazine; visits Lieutenant Powhatan Clarke at Fort Grant and accompanies him on a two-week scout on horseback with the Buffalo Soldiers among the Apaches near San Carlos in eastern Arizona. Remington befriends Lieutenant Carter Johnson of the 10th Cavalry and hears of his exploits with the Cheyenne.[76]
1896-09-01
Remington travels west as a guest of Lieutenant Carter Johnson, writing that he is going to Montana "to sweat & stink and thirst and starve and paint—particularly paint." He talks with Johnson about his exploits with the Northern Cheyenne seventeen years earlier, and makes the following note in his sketchbook: "The attack on the camp—Johnson charging

---

[74] John Arthur Baldwin biography at http://www.arlingtoncemetery.net/jabaldwin.htm (last visited May 28, 2018).

[75] Geoffrey Baldwin biography at http://www.arlingtoncemetery.net/gpbaldwin.htm (last visited May 29, 2018).

[76] Remington Timeline, Amon Carter Museum of Fine Art at http://www.cartermuseum.org/remington-and-russell/timeline?artist=All&narrative=All&page=2 (last visited May 28, 2018).

the infantry up the hill;" this becomes the inspiration for his oil, Through the Smoke Sprang the Daring Soldier (ACM), painted the following year.

1897-07-27

Remington receives a letter from Lieutenant Carter Johnson at Fort Assiniboine, Montana, praising him for the depiction of his exploits during the Cheyenne uprising as detailed in Remington's story, "The Sergeant of the Orphan Troop," to be published in Harper's New Monthly in August; one of the illustrations for the article is the painting Through the Smoke Sprang the Daring Soldier (ACM), and Johnson calls it "very true to life.[77]

The life of 1st Lt. Johnson was colorful, and a miniature of him based on the Remington painting serves as the symbol for the Southern California Historical Miniature Society.[78] His career after Arizona took him to Cuba and the Philippines. In 1909, he returned from the Philippines and then took an assignment as head of Fort Robinson in Nebraska, where he died a year later. The Carter P Johnson Lake is on the grounds of Fort Robinson, Nebraska.[79]

March 27, 1891, 1st Lt. Johnson and 2d Lt. Hughes were cited for "specially meritorious acts service" for "Vigilance and zeal, rapidity and persistency of pursuit, resulting in the capture of the renegade Indian Scouts from the San Carlos Agency and in making the surrounding country practically untenable for hostile Indians."

GENERAL ORDERS, }   HEADQUARTERS OF THE ARMY,
No. 34. }   ADJUTANT GENERAL'S OFFICE,
   Washington, March 27, 1891.

The Major General Commanding takes pleasure in publishing in orders to the Army the names of the following officers and enlisted men who, during the year 1887, distinguished themselves by "specially meritorious acts or conduct in service:"

January, 1887. Private *Henry S. Corp*, Company B, 4th Infantry (then of Troop L, 5th Cavalry): For skill, determination, and courage in encounter with trespassers within Cherokee Outlet, near Belter Creek, causing them to surrender.

June, 1887. Lieutenant Colonel *Henry W. Lawton*, inspector general (then captain, 4th Cavalry); Captain *Theodore J. Wint*, 4th Cavalry; 1st Lieutenant (then 2d lieutenant) *Carter P. Johnson*, 10th Cavalry; 2d Lieutenant *James B. Hughes*, 10th Cavalry: For vigilance and zeal, rapidity and persistency of pursuit, resulting in the capture of the renegade Indian scouts from the San Carlos Agency, and in making the surrounding country practically untenable for hostile Indians.

---

[77] Remington—Calvary Subjects, Amon Carter Museum of Fine Art at http://www.cartermuseum.org/timeline/narrative/remington-cavalry-subjects (last visited May 28, 2018),

[78] Little troops, Blog, "Quite a character – Carter P. Johnson," (Aug 4, 2011) at https://littletroops.wordpress.com/2011/08/04/quite-a-character-%E2%80%93-carter-p-johnson/ (last visited May 28, 2018).

[79] Ibid.

Capt. Alpheus H. Bowman 9th Infantry. In 1862, Capt. Bowman was himself the object of a court martial. He was dishonorable discharged from the U.S. Army with "conduct unbecoming an officer" after being provoked into a brawl with his 1st Lt.. He was later re-instated into the service by the Secretary of War. He went on the serve on other courts martial after the Apache Kid case.

Capt. T.E. Pierce was recognized in the Congressional Record for service with "Apache ... the most difficult in the world ...true savage," upon his retirement, March 31, 1888.

1892.   CONGRESSIONAL RECORD—SENATE.   2957

Let us see as to others in Dakota. Capt. James M. Bell was on duty at Pine Ridge Agency from May 22, 1886, to October 14, 1886. He was followed by Capt. Pierce in the same agency in Dakota from January 6, 1891, to February 5, 1891; he by Capt. Charles G. Penney, of whom we have heard, and he by Capt. George Le Roy Brown, who is now on duty at that agency.

At Rosebud, Capt. J. M. Lee performed duty as acting Indian agent from January 12, 1891, to April 23, 1891, and Capt. Cyrus A. Earnest performed similar duty from April 23, 1891, to July 20, 1891.

At Cheyenne River also in South Dakota, Capt. Joseph H. Hurst, of the Twelfth Infantry, was on duty from July 12, 1891, to October 4, 1891.

True it is that in the cases last mentioned, at the Rosebud and at the Cheyenne River Agency, these army officers were placed upon this duty as acting Indian agents because of emergency, because the Indian war was then on, and it was deemed best that they should perform this duty; but in the other instances the army officers were placed upon duty, not because an Indian war was on, but because it was thought best for the Indians, best for the Government, best for a proper administration of public affairs, that these gentlemen should act as agents for Indians. So that instead of one army officer only being thus detailed in South Dakota, we find that eleven have been in sixteen years.

In this list of thirty-five officers who have been thus detailed there are others who have received the commendation and praise of the Interior Department. I will refer now to an officer whose name I will have to use again when we return not to our mutton but to our bacon, for I must make a few more suggestions upon that odorous and odoriferous subject before I get through.

Capt. J. M. Lee, of the Ninth Infantry, was on duty at the Spotted Tail Agency in Nebraska, and he accompanied the Indians to their new agency on the Missouri River and the Secretary of the Interior requested the continuance of this army officer on duty with that agency.

Capt. A. R. Chaffee, of the Sixth Cavalry, was on duty at the San Carlos Agency in Arizona. The Indians at that agency were Apaches, the most difficult in the world to control, the nearest approach to the pure savage, perhaps, that there is upon this continent. When he left and was succeeded by a civilian agent he had been on duty there for nearly four years, and the Department of the Interior says it "regrets to lose the valuable services which have been exceptionally satisfactory to the Department."

Capt. Jno. L. Bullis, of the Twenty-fourth Infantry, was on duty at the same agency from April 23, 1888, to October 20, 1891, a number of years. The Interior Department said "this officer has been most efficient, both as officer and as agent. He had done his duty most faithfully."

Capt. Lee afterwards was placed on duty from 1885 to 1887 at the Cheyenne and Arapahoe Agency in the Indian Territory, and March 8, 1886, the Department expressed itself as "highly pleased with the manner in which Lieut. Lee had conducted all the business of the agency, with the progress made by the Indians, and with the peaceable condition of affairs since he assumed charge."

The tabular statement is as follows:

Statement of officers of the United States Army detailed for duty as Indian agents (presumably at the request of the Interior Department) from July 27, 1876, to March 1, 1892.

Name of officer, rank, and regiment.	Agency.	Tribe of Indians.	From—	To—
Oscar Elting, first lieutenant, Third Cavalry	Red Cloud, Nebr	Sioux	July 27, 1876	Sept. 20, 1876
Thomas F. Tobey, captain, Fourteenth Infantry	do	do	Sept. 20, 1876	Jan. 11, 1877
C. A. Johnson, first lieutenant, Fourteenth Infantry	do	do	Jan. 11, 1877	July 1, 1877
Do	do	do	Sept. 15, 1877	Oct. 26, 1877
John Bannister, first lieutenant, Twentieth Infantry	Standing Rock, Dak	do	Aug. 30, 1876	Sept. 16, 1876
R. E. Johnston, captain, First Infantry	do	do	Sept. 16, 1876	Dec. 9, 1876
Morris C. Foot, first lieutenant, Ninth Infantry	Spotted Tail, Nebr	do	Aug. 30, 1876	Oct. 27, 1876
A. C. Paul, first lieutenant, Third Cavalry	do	do	Oct. 27, 1876	Nov. 30, 1876
Horace Neide, first lieutenant, Fourth Cavalry	do	do	Nov. 30, 1876	Mar. 3, 1877
J. M. Lee, first lieutenant, Ninth Infantry a	do	do	Mar. 3, 1877	July 1, 1878
Richard C. Parker, captain, Twelfth Infantry	Hoopa Valley, Cal	Hoopas	July 9, 1877	Oct. 22, 1878
Henry R. Mizner, major, Eighth Infantry	do	do	Oct. 22, 1878	July 31, 1880
E. B. Savage, captain, Eighth Infantry	do	do	July 31, 1880	Mar. 1, 1881
Gordon Winslow, first lieutenant, Eighth Infantry	do	do	Mar. 1, 1881	Aug. 1, 1882
Charles Porter, captain, Eighth Infantry	do	do	Aug. 1, 1882	Aug. 6, 1885
J. N. Andrews, captain, Eighth Infantry	do	do	Aug. 6, 1885	July 1, 1886
William E. Dougherty, captain, First Infantry	do	do	July 1, 1886	Sept. 17, 1890
Do. b	Crow Creek and Lower Brule, S. Dak	Sioux	Mar. 12, 1878	Mar. 30, 1881
Theodore Schwan, captain, Eleventh Infantry c	Cheyenne River, Dak	do	Mar. 22, 1878	July 23, 1880
A. R. Chaffee, captain, Sixth Cavalry d	San Carlos, Ariz	Apache	July —, 1879	1880
T. E. Pierce, captain, First Infantry	do	do	Aug. 15, 1885	Mar. 31, 1888
Thaddeus W. Jones, first lieutenant, Tenth Cavalry	do	do	Apr. 5, 1885	Apr. 23, 1888
John W. Bullis, captain, Twenty-fourth Infantry e	do	do	Apr. 23, 1881	Oct. 20, 1891
Lewis Johnson, captain, Twenty-fourth Infantry	do	do	Oct. 20, 1891	(f)
F. T. Burnett, captain, Ninth Cavalry	Navajo, Ariz	Navajo	June 12, 1880	July 11, 1881
Elias Chandler, second lieutenant, Sixteenth Infantry	Tonkawas, Tex	Tonkawa	Jan. 2, 1882	June 29, 1885
J. M. Lee, first lieutenant, Ninth Infantry g	Cheyenne and Arapahoe, Ind. T	Cheyenne and Arapahoe	July 27, 1885	Sept. 16, 1887
J. M. Bell, captain, Seventh Cavalry	Pine Ridge, Dak	Sioux	May 22, 1886	Oct. 14, 1886
F. J. Pierce, captain, First Infantry	do	do	Jan. 6, 1891	Feb. 5, 1891
Charles G. Penney, captain, Sixth Infantry h	do	do	Feb. 5, 1891	Oct. 30, 1891
George Le R. Brown, captain, Eleventh Infantry	do	do	Oct. 27, 1891	
Carroll H. Potter, captain, Eighteenth Infantry	Osage, Ind. T	Osage	May 16, 1887	Aug. 5, 1888
J. M. Lee, first lieutenant, Ninth Infantry	Rosebud, S. Dak	Sioux	Jan. 12, 1891	Apr. 23, 1891
Cyrus A. Earnest, captain, First Infantry	do	do	Apr. 23, 1891	July 30, 1891
Joseph H. Hurst, captain, Twelfth Infantry	Cheyenne River, S. Dak	do	Jan. 12, 1891	Oct. 21, 1891
E. P. Ewers, captain, Fifth Infantry	Tongue River, Mont	Northern Cheyenne	Jan. 12, 1891	Nov. 28, 1891

Remarks by Interior Department when relieved (if any).

a Lieut. Lee accompanied the Indians to their new agency on the Missouri River, and the Secretary of the Interior requested the continuance of Lieut. Lee on duty with the agency.

b Secretary of the Interior thanks Capt. Dougherty for his valuable services, rendered the Department under peculiar circumstances; also for his uniform courtesy and prompt obedience to orders from the Department.

c "It is with extreme reluctance that a change at the agency has been acceded to; the thorough and able administration of agency affairs is highly appreciated by the Interior Department and the thanks of the Department are extended for the peculiarly difficult and arduous services which have been rendered."

Capt. Bullis who pursued Apache Kid in the territorial criminal trial in 1889, was retired October 20, 1891, only a few months after he fired Al Sieber.[80]

Gen. Nelson Miles served from 1895 to 1903. He served as the last Commanding General of the United States Army during the Spanish-American War. He was involved in almost every major American battle against American Indians, slaughtering untold numbers of American Indians. He was on the ballot for Presidential candidate at the 1904 Democrat National Convention. He died from a heart attack at the circus with his grandchildren on May 16, 1925.[81]

Acting Judge Advocate General Brig. Gen. G. Norman Lieber became the Judge Advocate General (1895-1901).

His letter of appeal, April 11, 1888, made clear that he knew the integrity of the process was at issue as well as the fate of the defendants. He was the son of the former Judge Advocate General, Francis Norman Lieber, author of the Lieber Code concerning how to conduct war.

Final thoughts

This case of the Apache Kid shows how one unusually talented individual bridged his culture to that of the colonizers in western culture and the military culture within that world, and had to navigate the judicial branch of each one of these worlds. His demonstrated remarkable tenacity and understanding in following the complex procedural and substantive rules of the military justice system, the federal justice system and the territorial justice system, while fulfilling his legal and cultural obligations in the Apache judicial system. Add to this a language barrier. This might be difficult enough, but this was in the context of the decimation of his own people, through killings, executions or removals, which alone was an all-consuming trauma. It was remarkable how his defense counsel, 1st Lt. Baldwin was able to articulate the hatred toward the Apaches alongside the "progress" he attributed to the Scouts in assimilating with the Army to his military court members, albeit falling to deaf ears. His closing argument is a valuable insight into the better thinking of some of the military officers with regard to the Apache people.

This case represents the end of an era in American history, and the end of one of the last truly independent Indian nations, destined to become one of the "domestic dependent nations"[82] under the U.S. Constitution, like other tribes. The value in an ethnohistory using legal documents

---

[80] 23 Cong. Rec. 2957 (1892).

[81] Paul Andrew Hutton, *The Apache Wars* 422 (2012).

[82] *Cherokee Nation v. Georgia* (30 U.S. (5 Pet.) 1 (1831)).

like this provides some framework for the actions of the dominant legal society and how all parties respond and navigate their lives.

Understanding tribal culture is critical to understanding motives and whether there are actually crimes that have been committed. Al Sieber was said to have remarked that, "Kid's downfall was more misunderstanding than criminal intention."[83] Sieber's actions did not show such benevolence if he really believed what he said.

How much have we improved our understanding of culture in the courtroom over more than a century since this case? In 2010, former Military Judge Daniel Benson and Professor of Law at Texas Tech University, described a case over which he presided as sentencing judge after a court martial at Ft. Bliss, Texas. He said the case involved a "young Apache soldier" who was absent without leave for two weeks. He was found guilty of AWOL, and was sent to Judge Benson's court for sentencing and he learned that the soldier had decided to plead guilty. Judge Benson called on an expert anthropologist to provide evidence in the sentencing hearing about why he was absent. It turns out that the soldier was absent in order to fulfil a tribal obligation in what would be a "coming of age" ceremony for men. He had to go at that particular time because in part, he was engaged to marry, and he would not be allowed to marry without having first gone through this ceremony. Judge Benson gave him a lenient sentence of a short time without pay, which is a minimal punishment.[84]

This Apache soldier of the 20th Century really had no choice, sadly, a century after the Apache Kid case, because he, too, knew it was hopeless to try to explain how important this obligation was to him and chose the consequences of violating military law.

---

[83] Dan R. Williamson, "Apache Kid: Red Renegade of the West," 31 *Arizona Highways* 15 (May 1939).

[84] Digital recording of a Panel, The Court Martial of Apache Kid, at Texas Tech University School of Law, Nov. 1, 2010 (in the possession of the author).

www.ingramcontent.com/pod-product-compliance
Lightning Source LLC
Chambersburg PA
CBHW051337200326
41519CB00026B/7452